STEWART HARRIS (1937–2004) was a faculty member in the Department of Mechanical Engineering at Stony Brook University from 1966 until his death. He served as chairman of the department from 1978 to 1981 and was the dean of the College of Engineering from 1981 to 1992. Professor Harris held many visiting appointments in the U.S. and in Europe; these included positions at Princeton University, Imperial College (London), the Materials Directorate at Wright Patterson Air Force Base, and Columbia University.

Dr. Harris received a B.S.E. from Case Institute of Technology (1959) and an M.S. (1960) and Ph.D. (1964) from Northwestern University, followed by postdoctoral work at the Courant Institute from 1964 to 1966. His research focused on the theory of transport processes at the microscopic level and on molecular beam epitaxy—the growth of atomically precise metal and semiconductor crystal structures. Dr. Harris's research interests encompassed a number of distinct areas related to the study of reaction-diffusion systems; most recently these included the dispersal of genes and organisms, including early human populations.

An Introduction to the Theory of the BOLTZMANN EQUATION

Stewart Harris
State University of New York at Stony Brook

DOVER PUBLICATIONS
Garden City, New York

Bibliographical Note

This Dover edition, first published in 2004, is an unabridged republication of the work originally published by Holt, Rinehart and Winston, Inc., New York, in 1971.

Library of Congress Cataloging-in-Publication Data

Harris, Stewart.
An introduction to the theory of the Boltzmann equation / Stewart Harris.
p. cm.
Originally published: New York : Holt, Rinehart, and Winston, 1971.
Includes bibliographical references and index.
ISBN-13: 978-0-486-43831-3
ISBN-10: 0-486-43831-7
1. Transport theory. I. Title.

QC175.2.H33 2004
530.13'8—dc22

2004050198

Manufactured in the United States of America
43831707 2022
www.doverpublications.com

To My Parents, Seymour and Goldie
And My Wife, Helen

Faith is a fine invention
For gentlemen who see;
But microscopes are prudent
In an emergency!

Emily Dickinson

Preface

The past twenty years have seen the steady maturation of the theory of the Boltzmann equation. In this period the Boltzmann equation has gone from a rather formidable mathematical curiosity with a limited but distinguished following, to a much used methodology having widespread application. With the advent of modern high-speed flight through the rarefied regions of the outer atmosphere Boltzmann's equation has taken on a particular practical significance. However, it is no longer the exclusive property of the fluid physicist. One finds Boltzmann's equation, or Boltzmann-like equations, extensively used in such disparate fields as laser scattering, solid-state physics, nuclear transport, and even outside the conventional boundaries of physics and engineering in the fields of cellular proliferation and automobile traffic flow. It is especially significant that at a time when the method of time correlation functions seemed to be usurping much of the Boltzmann equation's earlier domain of transport theory such a veritable cornucopia of new uses for this equation should arise.

Many of the advances which have been made in the theory of Boltzmann's equation since the publishing, some thirty years ago, of Chapman and Cowling's now classic monograph are contained in the *Handbuch* article of Grad, who is himself responsible for much of this work. There exist two other outstanding expositions on the theory of the Boltzmann equation, the book by Carleman, and the *Handbuch* article by Waldmann—neither, unfortunately, are at present available in English. The above-mentioned sources are intended for a relatively sophisticated audience and for that reason are not basically suitable for use as a text in an introductory graduate level course on the theory of Boltzmann's equation; in fact, strangely enough, there seems to be no book at the present time which satisfies this need. (Shorter, less complete accounts of the theory of Boltzmann's equation appear as an isolated chapter or two in books mainly devoted to equilibrium statistical mechanics.) This situation has been evident to me for some time, both as a graduate student and as a teacher of graduate students, and I have finally become emboldened enough to attempt to remedy what I consider to be a serious void in textbook literature.

In this book I have attempted to present in some detail the basic modern theory of Boltzmann's equation, and to include representative applications using both Boltzmann's equation and the model Boltzmann equations which are developed in the text. It is hoped that adherence to detail will enable this book to also serve as a reference source for research workers using the Boltzmann equation. The book is primarily intended for use in a graduate level course on the theory of the Boltzmann equation for physicists and engineers, and to this end I have tended to emphasize the physical aspects of the theory. The problems following each chapter are intended as learning examples, and these have quite often been used to extend and generalize the text material. This book has been the direct product of a course on the "Theory of the Boltzmann Equation" which I have taught at Stony Brook for the past few years. The point of view represents a distillation and, hopefully, a coherent synthesis of ideas presented to me from several sources. In particular, courses by Professors I. M. Krieger, M. B. Lewis, I. Prigogine, and G. E. Uhlenbeck, and the books by Chapman and Cowling, and by Carleman have been especially stimulating. A special debt must be acknowledged to Professor H. Grad who, through both the spoken and written word, has taught me much of what I know about the Boltzmann equation. Any successes which this book may enjoy are due in part to the above men; its failures must be my own.

Stony Brook, New York
April 1971

STEWART HARRIS

Historical Introduction

In the Foreword to Part 2 of his *Lectures on Gas Theory*, written in 1898, Boltzmann comments on the increasing attack which was being mounted by the so-called "school of energetics" on all theories based on the atomic model of matter. It is unfortunate that it was only after Boltzmann's untimely death, eight years later, that the atomic theory of matter became universally accepted by the scientific community, and the twentieth century began for physics and chemistry.

The idea that matter is composed of atoms arose to conflict with the earth, water, air, and fire theory sometime in the fifth century B.C., and is usually credited to the Greek philosophers Leucippus and Democritus. Although discredited by Aristotle, the idea survived somehow until the first century A.D., when it was championed by the Roman poet–philosopher Lucretius. In the Middle Ages the ideas of Aristotle became firmly established as Church dogma, and atomism was relegated to the list of heresies. For this, among other heresies, the ex-monk and philosopher Giordano Bruno fell victim to the flames of the Inquisition in 1600. Ideas cannot be killed as easily as men, however, and in the seventeenth century, in a somewhat more tolerant atmosphere, the atomic theory was again taken up by Gassendi, and then by Boyle and Newton. Finally, in 1738, Daniel Bernoulli used the atomic model as the basis for postulating the first kinetic theory of gases. After another century of inactivity, Herapath, Waterson, Clausius, and finally the masters, Maxwell and Boltzmann, formulated and developed a rigorous kinetic theory of gases. Herapath and Waterson found their ideas rejected by the scientific establishment of their country (they were English), although Waterson's paper on the subject was finally published in 1891 by Lord Rayleigh, eight years after Waterson's death and forty-five years after he first submitted it to the Royal Society. Clausius, Maxwell, and Boltzmann, however, were important scientists in their countries, and although they were swimming against the scientific current of their day, they were able to publish their results. The work of Maxwell and Boltzmann, especially, can be considered as forming the basis of the modern kinetic theory of rarefied gases which we will consider in this book.

Contents

An Introduction to the Theory of the
BOLTZMANN EQUATION

CHAPTER 1

statistical mechanical preliminaries

1-1 THE MICROSCOPIC DESCRIPTION

Equilibrium and nonequilibrium statistical mechanics have as one of their objectives the determination of the basic thermo-fluid properties of systems which are characterized by having a large number of degrees of freedom. The description is on the molecular level and is usually given in terms of the spatial coordinate and momentum of each of the molecules that constitute the system of interest. This is to be contrasted with the corresponding thermodynamic description, which for a one-component system requires but three variables, only two of which are independent. For a fluid mechanical description we must specify in addition to the three thermodynamic variables the velocity field, stress tensor, and heat flux vector. The last two quantities are not usually treated as independent variables, however, so that a minimal fluid mechanical description requires five independent variables. The statistical mechanical description of the same system requires $6N$ variables, where N is the number of molecules in the system, typically about 10^{23}. As mentioned above, these variables are usually taken to be the $3N$ spatial coordinates, $\mathbf{q}_1, \mathbf{q}_2, \cdots, \mathbf{q}_N \equiv \{\mathbf{q}\}$, and the $3N$ conjugate momenta, $\mathbf{p}_1, \mathbf{p}_2, \cdots \mathbf{p}_N \equiv \{\mathbf{p}\}$, of the constituent molecules. Here we have already simplified matters by considering the spatial specification of a molecule to be completely determined by a point in space, so that we have neglected orientation effects, vibrations, and, in fact, all phenomena which are related to the structure of the molecules. We will generally restrict ourselves to structureless particles throughout this book.

The equations of motion of the molecular system we are considering are given most conveniently in terms of the Hamiltonian of this system, which we denote as $H(\{\mathbf{q}\},\{\mathbf{p}\})$, where

$$H(\{\mathbf{q}\},\{\mathbf{p}\}) = E(\{\mathbf{q}\},\{\mathbf{p}\}) \tag{1-1}$$

In the absence of external fields, $E(\{\mathbf{q}\},\{\mathbf{p}\})$ is the total energy, kinetic energy plus any potential energy due to intermolecular forces, of the system. In terms of H, the Hamiltonian formulation of the equations of motion for the system is

$$\dot{\mathbf{q}}_i = \frac{\partial H}{\partial \mathbf{p}_i} \qquad i = 1, 2, \cdots, N$$

$$\dot{\mathbf{p}}_i = \frac{-\partial H}{\partial \mathbf{q}_i} \tag{1-2}$$

The state of the system at any time can be specified by a single point in the $6N$ dimensional space (called the gamma space, or Γ space) having the mutually orthogonal axis $\mathbf{q}_1, \mathbf{q}_2, \cdots, \mathbf{p}_{N-1}, \mathbf{p}_N$. This space is an example of what is more generally called a phase space; a point in the phase space of a particular system determines the phase, that is, state, of that system. If we know what the intermolecular forces are, then given any point in the gamma space (initial data), we can determine from Equation (1–2), in theory, the state of the system at any later or earlier time. Thus Equation (1–2) provides us with the *modus operandi* for calculating the trajectories of a particular system in the phase space, provided we are supplied with some initial conditions.

1–2 RELATIONSHIP BETWEEN MICROSCOPIC AND MACROSCOPIC DESCRIPTIONS

We will accept the following statement as a basic postulate of both equilibrium and nonequilibrium statistical mechanics.

Postulate

All intensive macroscopic properties of a given system can be described in terms of the microscopic state of that system.

Thus any intensive macroscopic property which we may observe by measurement, G_{obs}, can be also written as $G(\{\mathbf{q}\},\{\mathbf{p}\})$ and determined directly from a knowledge of the microscopic state. In most cases it may be a difficult task to determine the proper function G, but it is nevertheless postulated that for every intensive macroscopic property there exists a G. For the common thermo-fluid macroscopic properties we shall consider here we will always be able to explicitly display the corresponding G.

In measuring any macroscopic quantity the associated physical measurement is never instantaneous, but rather must always be carried out over some finite time τ. This interval may be made very small by refinement of observational technique, but it will always remain finite. The quantity we observe is thus not the instantaneous value, $G(\{\mathbf{q}\},\{\mathbf{p}\})$, but rather a time-averaged value of G, $\bar{G}$, where if t_0 is the time the observation is begun

$$\bar{G}(\tau) = G_{\text{obs}} = \frac{1}{\tau}\int_{t_0}^{t_0+\tau} G(\{\mathbf{q}(t)\},\{\mathbf{p}(t)\})\,dt \qquad (1\text{-}3)$$

If the proper value of G associated with a particular macroscopic quantity is known, then the observed value can be calculated from Equation (1–3), using the equations of motion given by Equation (1–2), provided that the microscopic state of the system is known at some given time. In practice such a calculation is, of course, impossible. We are not able to determine the microscopic state at any given time, and even were such a specification possible, we would be faced with an overwhelming number of equations to solve. Thus any attempt to put Equation (1–3) to practical use will involve an insurmountable computational problem, due primarily to the large number of degrees of freedom of the systems which we typically encounter.

1–3 THE GIBBS ENSEMBLE

Due to the difficulties we may expect to occur in attempting to directly calculate $\bar{G}$ from Equation (1–3), we turn to another method of determining G_{obs}. This is the method of ensembles, originally due to Gibbs. We consider in the "minds eye" a large number of systems, η, which macroscopically are equivalent to the system we are considering. The thermo-fluid properties of each of the η replicas are then equal to those of the system of interest. The microscopic description is not specified, and we would expect that this will differ greatly among the replica systems in the absence of such specification, since there are large numbers of microscopic states corresponding to any given macroscopic state. This collection of systems is referred to as an ensemble.

Each system in the ensemble can be represented by a point in the phase space. If we let $\eta \to \infty$ these points become quite dense in the phase space, and we can describe their distribution throughout the phase space by a density function which is a continuous function of $\{\mathbf{q}\},\{\mathbf{p}\}$. We can also normalize this function so that it is a probability density in the phase space; the normalized probability density function will be denoted by $F_N(\{\mathbf{q}\},\{\mathbf{p}\},t) = F_N$. F_N will be symmetric in the $\mathbf{q}_i$ and $\mathbf{p}_i$.

Specifically we will have

$$F_N(\{\mathbf{q}\},\{\mathbf{p}\},t)\prod_{i=1}^{N} d\mathbf{q}_i\, d\mathbf{p}_i = \text{the fraction of the } \eta \text{ phase points which at time } t \text{ are in the incremental volume element } \prod_{i=1}^{N} d\mathbf{q}_i\, d\mathbf{p}_i \text{ about the point } \{\mathbf{q}\}, \{\mathbf{p}\} \text{ in the phase space} \tag{1-4}$$

Although F_N is itself a probability, it changes in time in a completely deterministic manner. If we know the dependence of F_N on the phase at a particular time, then by solving Hamilton's equations, Equation (1–2), we can determine F_N at any later (or earlier) time.

An ensemble average of the macroscopic property $G(\{\mathbf{q}\},\{\mathbf{p}\})$ can be defined in the following manner:

$$\langle G(t)\rangle = \int \prod_{i=1}^{N} d\mathbf{q}_i\, d\mathbf{p}_i G(\{\mathbf{q}\},\{\mathbf{p}\})F_N(\{\mathbf{q}\},\{\mathbf{p}\},t) \tag{1-5}$$

The second basic postulate of statistical mechanics which we shall make use of is the following, sometimes called the ergodic statement. (The proof that $\lim_{\tau\to\infty} \bar{G}(\tau)$ exists is also referred to as the ergodic statement.)

Postulate

$$\bar{G}(t) = \langle G(t)\rangle$$

The above postulate states that we can consider ensemble averages rather than time averages as a basis for determining macroscopic properties from the microscopic description. Thus we are led to study in detail the properties of the probability density F_N. As a final remark we note that the validity of the above postulates, here as elsewhere, must be ultimately measured in the absence of rigorous proof by how well the predictions based on them compare with experiment.

1–4 THE LIOUVILLE EQUATION

We are interested in determining how F_N evolves in time due to the natural motion of each ensemble member in the phase space. The Hamilton equations, Equation (1–2), determine how the phase of each system changes in time, and so we would expect that they will play a central role in determining the temporal behavior of F_N as well.

Let us consider the change, dF_N, in the value of F_N at the phase point $\{\mathbf{x}\} \equiv \{\mathbf{q}\},\{\mathbf{p}\}$ at the time t which results from arbitrary, infinitesimal changes in these variables. We will have

$$dF_N = \frac{\partial F_N}{\partial t}\,dt + \sum_{i=1}^{N} \frac{\partial F_N}{\partial \mathbf{q}_i} \cdot d\mathbf{q}_i + \sum_{i=1}^{N} \frac{\partial F_N}{\partial \mathbf{p}_i} \cdot d\mathbf{p}_i \tag{1-6}$$

This expression must be valid for all infinitesimal changes $\{\mathbf{x}\} \rightarrow \{\mathbf{x} + d\mathbf{x}\}$, so that we can specify that this infinitesimal change be just that which follows the trajectory of the system in the phase space. We then have $(d\mathbf{q}_i/dt) = \dot{\mathbf{q}}_i$, $(d\mathbf{p}_i/dt) = \dot{\mathbf{p}}_i$, and for this case Equation (1–6) can be rewritten as

$$\frac{dF_N}{dt} = \frac{\partial F_N}{\partial t} + \sum_{i=1}^{N} \left[\frac{\partial F_N}{\partial \mathbf{q}_i} \cdot \dot{\mathbf{q}}_i + \frac{\partial F_N}{\partial \mathbf{p}_i} \cdot \dot{\mathbf{p}}_i \right] \tag{1-7}$$

In the above equation $\partial F_N/\partial t$ is the local change of F_N, that is, the change at the point $\{\mathbf{x}\}$, and dF_N/dt is the total change of F_N along the trajectory in the neighborhood of $\{\mathbf{x}\}$. Liouville's theorem now states that

$$\frac{dF_N}{dt} = 0 \tag{1-8}$$

The statement of Liouville's theorem, Equation (1–8), is referred to as Liouville's equation. Before proving this result let us first consider its implications. In words, Liouville's theorem states that along the trajectory of any phase point the probability density in the neighborhood of this point remains constant in time. From this it follows that an incremental volume about the phase point $\mathbf{x}$, whose surface $S(\mathbf{x})$ is defined by a particular set of phase points, is also invariant in time. Thus as the point $\mathbf{x}$, moves along its trajectory, the volume $d\mathbf{x}$ defined by $S(\mathbf{x})$ changes in shape, but its total volume remains constant. This is easily seen as a consequence of the fact that Hamilton's equations have unique solutions, so that there can be no intersection of the trajectories of separate ensemble members in the phase space. The points inside $d\mathbf{x}$ can never cross the surface $S(\mathbf{x})$, then, since this would mean that their trajectories intersect those of the points which define $S(\mathbf{x})$. Since both F_N and the number of points inside $d\mathbf{x}$ remain constant in time, we can conclude that the volume of $d\mathbf{x}$ must also remain unchanged. A formal way of stating the above is to say that the transformation of the phase space into itself induced by the flow described by Hamilton's equations preserves volumes.

Since F_N remains constant along a trajectory in the phase space, any function of F_N will also have this property. Of particular interest will be the function

$$H(t) \equiv \int_\Gamma d\mathbf{x}\, F_N \ln F_N = \text{constant}$$

We will return to this function later when we discuss the ideas of reversibility and irreversibility. In this context we mention here that Liouville's equation is an example of a reversible equation. By this we mean that the transformation $t \to -t$ leaves the form of the equation unaltered so that if $F_N(\{\mathbf{q}(t)\},\{\mathbf{p}(t)\},t)$ is a solution so is $F_N(\{\mathbf{q}(-t)\},\{-\mathbf{p}(-t)\},-t)$.

1–5 PROOF OF LIOUVILLE'S THEOREM

We consider here a simple proof of Liouville's theorem. Consider an arbitrary but fixed volume V in the phase space, and let its surface be denoted as S. The number of phase points inside V at the time t is η_V, where

$$\eta_V = \eta \int_V d\mathbf{x}\, F_N$$

Since V is fixed, we have

$$\frac{d\eta_V}{dt} = \eta \int_V d\mathbf{x}\, \frac{\partial F_N}{\partial t} \tag{1–9}$$

We can also obtain an expression for $d\eta_V/dt$, the net rate of change in the number of phase points inside V, by equating this quantity to the net flow of phase points passing through S. If $\{\dot{\mathbf{x}}\}$ denotes the $6N$-dimensional flow vector in the phase space having coordinates $\dot{\mathbf{q}}_1, \dot{\mathbf{q}}_2, \cdots, \dot{\mathbf{p}}_{N-1}, \dot{\mathbf{p}}_N$ and $\hat{\mathbf{n}}$ is the unit normal at S, then we have

$$\frac{d\eta_V}{dt} = -\eta \int_S dS(\hat{\mathbf{n}} \cdot \dot{\mathbf{x}})F_N \tag{1–10}$$

The minus sign is due to the fact that $\hat{\mathbf{n}}$ "points out" of S. The surface integral in the above equation can be written as a volume integral by making use of a generalization of Gauss's theorem, so that

$$\frac{d\eta_V}{dt} = -\eta \int_V d\mathbf{x}\, \nabla_x \cdot (\dot{\mathbf{x}}F_N) \tag{1–11}$$

where ∇_x is the gradient operator in the phase space. Since V was arbitrarily chosen, we can combine Equations (1–9) and (1–11) to obtain the following result:

$$\frac{\partial F_N}{\partial t} + \nabla_x \cdot (\dot{\mathbf{x}}F_N) = 0$$

Writing out the gradient operator and flow vector we then have,

$$\frac{\partial F_N}{\partial t} + \sum_{i=1}^{N} \left(\frac{\partial F_N}{\partial \mathbf{p}_i} \cdot \dot{\mathbf{p}}_i + \frac{\partial F_N}{\partial \mathbf{q}_i} \cdot \dot{\mathbf{q}}_i \right) + \sum_{i=1}^{N} F_N \left(\frac{\partial \dot{\mathbf{p}}_i}{\partial \mathbf{p}} + \frac{\partial \dot{\mathbf{q}}_i}{\partial \mathbf{q}} \right) = 0 \quad (1\text{-}12)$$

From Hamilton's equations we see that the last term is identically zero:

$$\frac{\partial \dot{\mathbf{p}}_i}{\partial \mathbf{p}_i} + \frac{\partial \dot{\mathbf{q}}_i}{\partial \dot{\mathbf{q}}_i} = \frac{\partial^2 H}{\partial \mathbf{p}_i \, \partial \mathbf{q}_i} - \frac{\partial^2 H}{\partial \mathbf{q}_i \, \partial \mathbf{p}_i} = 0$$

so that Equation (1–12) reduces to Liouville's equation

$$\frac{\partial F_N}{\partial t} + \sum_{i=1}^{N} \left(\frac{\partial F_N}{\partial \mathbf{p}_i} \cdot \dot{\mathbf{p}}_i + \frac{\partial F_N}{\partial \mathbf{q}_i} \cdot \dot{\mathbf{q}}_i \right) = 0 \quad (1\text{-}13)$$

which completes our proof of Liouville's theorem.

At this point one might reasonably wonder why we have gone to all the trouble of introducing the Gibbs ensemble and ensemble density F_N. We are certainly still faced, in the presence of the Liouville equation, with an impossible computational problem due to the overwhelmingly large number of degrees of freedom present. This question will be answered in the next chapter, where we consider a more tractable description which we obtain directly from the ensemble formalism.

References

The standard reference on Hamiltonian mechanics is:

1. H. Goldstein, *Classical Mechanics*. Reading, Mass.: Addison-Wesley, 1959.

The statistical mechanical preliminaries are elaborated upon in:

2. T. L. Hill, *Statistical Mechanics*. New York: McGraw-Hill, 1956.
3. A. I. Khinchin, *Mathematical Foundations of Statistical Mechanics*. New York: Dover, 1949.
4. R. Tolman, *The Principles of Statistical Mechanics*. Oxford, England: Oxford University Press, 1938.
5. G. E. Uhlenbeck and G. Ford, *Lectures in Statistical Mechanics*. Providence, R.I.: American Mathematical Society, 1963.

6. R. L. Liboff, *Theory of Kinetic Equations*. New York: Wiley, 1968.

Problems

1–1. The cannonical distribution function of equilibrium statistical mechanics is $F_N = Ae^{-\beta H}$, where A and β are constants and H is the Hamiltonian of the system being described. Determine the constant A in terms of β and H.

1–2. Show that $F_N = Ae^{-\beta H}$ is a stationary solution of Liouville's equation (A, β, and H are defined in Problem 1–1).

1–3. Consider the trajectory of a specified system in the gamma space. On this trajectory we label points $\Gamma_0, \Gamma_1, \cdots, \Gamma_n$ with $n \to \infty$ in such a way that it takes the system the same amount of time, say τ, to move from each Γ_i to Γ_{i+1}. We can compute the time average, $\bar{G}(t)$, for any function G of the phase coordinates on this trajectory. By replacing the integral which appears in the time average by a sum, show that this time average is equal to the ensemble average when the ensemble is prepared so that the states of the ensemble members are given by the Γ_i.

1–4. Prove that the solution to Hamilton's equations are unique for a given Hamiltonian; that is, if two trajectories in the gamma space pass through $\Gamma(t + \Delta t)$ corresponding to the state of the system at time $t + \Delta t$, then they both must have originated from the same $\Gamma(t)$.

1–5. An incremental volume in the gamma space, $d\Gamma$, is transformed into the volume $d\Gamma'$ through a flow described by Hamilton's equations. As discussed in Section 1–4, $d\Gamma = d\Gamma'$. Prove this for a system of noninteracting particles by showing that the Jacobian of the transformation is unity.

1–6. Write Liouville's equation for a system which is acted on by a time-independent external field.

1–7. The Liouville equation can be written as

$$\frac{\partial F_N}{\partial t} + H_N F_N = 0$$

(what is the operator H_N?) so that a formal solution of this equation is

$$F_N(\{\mathbf{q}\},\{\mathbf{p}\},t) = e^{-tH_N} F_N(\{\mathbf{q}\},\{\mathbf{p}\},0)$$

where

$$e^{-tH_N} \equiv \sum_n (-tH_N)^n/n!$$

Prove that for any function of the phase variables G,

$$e^{-tH_N}G(\{\mathbf{q}\},\{\mathbf{p}\}) = G(\{e^{-tH_N}\mathbf{q}\},\{e^{-tH_N}\mathbf{p}\})$$

and interpret the effect of e^{-tH_N} acting on G.

1–8. Using the results of Problems 1–6 and 1–7, write Liouville's equation in the form

$$\frac{\partial F_N}{\partial t} + (H_N + H_N')F_N = 0$$

where H_N' is the contribution of a time-independent external field. Show that a solution to the above equation is given by

$$F_N(\{\mathbf{q}\},\{\mathbf{p}\},t) = e^{-tH_N}F_N(\{\mathbf{q}\},\{\mathbf{p}\},t) - \int_0^t dS\, e^{-(t-S)(H_N+H_N')}H_N'F_N(\{\mathbf{q}\},\{\mathbf{p}\},0)$$

1–9. Show that solving Liouville's equation for a given Hamiltonian is equivalent to solving the equations of motion for the dynamical system, and do this explicitly for a system of non-interacting particles.

CHAPTER 2

contraction of the statistical mechanical description

2-1 REDUCED DISTRIBUTION FUNCTIONS

As we have seen in the preceding chapter, any attempt to consider a given system in the context of the full statistical mechanical description invariably leads to an intractable computational problem due to the large number of degrees of freedom which appear. Fortunately the macroscopic properties of greatest interest do not depend on averages taken with respect to F_N, but rather, as we shall see shortly, on averages taken with respect to the first few so-called reduced distribution functions. Let us define the R particle reduced distribution function, or R particle distribution function, $F_R(\mathbf{x}_1,\mathbf{x}_2, \cdots, \mathbf{x}_R,t)$, in the following way:[1]

$$F_R(\mathbf{x}_1,\mathbf{x}_2, \cdots ,\mathbf{x}_R,t) \equiv \int d\mathbf{x}_{R+1} \cdots d\mathbf{x}_N\, F_N(\mathbf{x}_1, \cdots ,\mathbf{x}_R, \cdots ,\mathbf{x}_N,t) \tag{2-1}$$

As we mentioned above, our interest will be particularly focused on the first few reduced distribution functions:

$$F_1(\mathbf{x}_1,t) = \int d\mathbf{x}_2 \cdots d\mathbf{x}_N\, F_N(\mathbf{x}_1, \cdots ,\mathbf{x}_N,t) \tag{2-2}$$

and

$$F_2(\mathbf{x}_1,\mathbf{x}_2,t) = \int d\mathbf{x}_3 \cdots d\mathbf{x}_N\, F_N(\mathbf{x}_1, \cdots ,\mathbf{x}_N,t) \tag{2-3}$$

From our earlier definition of F_N it should be clear that $F_1(\mathbf{x}_1,t)\, d\mathbf{x}_1$ is the probability of finding molecule 1 in the incremental volume element $d\mathbf{x}_1$ about the phase point $\mathbf{x}_1$ at time t. Further, we can define a number density, $f_N(\mathbf{x}_1,t) = NF_1(\mathbf{x}_1,t)$, and a mass density, $f(\mathbf{x}_1,t) = NmF_1(\mathbf{x}_1,t)$ (m = the mass of a single molecule), such that $f_N(\mathbf{x}_1,t)\, d\mathbf{x}_1$ and $f(\mathbf{x}_1,t)\, d\mathbf{x}_1$ are the expected number of molecules and the expected mass, respectively,

[1] Unless we specify differently in the limits of integration $\int d\mathbf{x}_R \cdots d\mathbf{x}_{R+S}$ will always mean

$$\int_{\text{all } \mathbf{x}_R} d\mathbf{x}_R \int_{\text{all } \mathbf{x}_{R+1}} d\mathbf{x}_{R+1} \cdots \int_{\text{all } \mathbf{x}_{R+S}} d\mathbf{x}_{R+S}$$

within the incremental volume element $d\mathbf{x}_1$ about the phase point $\mathbf{x}_1$ at time t.

The F_R will be symmetric in their R arguments $\mathbf{x}_1, \mathbf{x}_2, \cdots, \mathbf{x}_R$. This property results from the fact that, as we have specified earlier, it is valid for the special case where $R = N$.

In order to describe the behavior of the F_R we need an equation which gives their temporal development. To obtain such an equation we start with the Liouville equation, which it is convenient to rewrite in the following form:

$$\frac{\partial F_N}{\partial t} + \sum_{i=1}^{N} \frac{\mathbf{p}_i}{m} \cdot \frac{\partial F_N}{\partial \mathbf{q}_i} + \sum_{i=1}^{N} \mathbf{F}_i \cdot \frac{\partial F_N}{\partial \mathbf{p}_i} = 0 \tag{2-4}$$

Here we have identified $\dot{\mathbf{q}}_i = \mathbf{p}_i/m$ and $\dot{\mathbf{p}}_i = \mathbf{F}_i$, where $\mathbf{F}_i$ is the net force which acts on the ith molecule. For convenience we will consider the case where no external force is acting on the system so that $\mathbf{F}_i$ is only due to the force exerted on molecule i by the other molecules of the system. In a statistical mechanical description it is usually assumed that the forces between the particles are derivable from a two-particle potential, so that the force exerted on the ith molecule by the jth is given from a potential, ϕ_{ij}, which is a function of $|\mathbf{r}_{ij}| \equiv |\mathbf{q}_i - \mathbf{q}_j|$. Making this assumption, we then have

$$\mathbf{F}_i = -\sum_{j=1 \neq i}^{N} \frac{\partial \phi_{ij}}{\partial \mathbf{q}_i} \tag{2-5}$$

In Chapter 3 we will consider the subject of intermolecular forces in more detail.

An equation for the F_R can be obtained from Equation (2–4) by integrating this equation over the phases $\mathbf{x}_{R+1}, \cdots, \mathbf{x}_N$. Let us consider each term in this integrated equation separately. The first term can be simplified by making use of the fact that the limits of integration are fixed, so that making use of the definition of Equation (2–1) we have

$$\int d\mathbf{x}_{R+1} \cdots d\mathbf{x}_N \frac{\partial F_N}{\partial t} = \frac{\partial}{\partial t} \int d\mathbf{x}_{R+1} \cdots d\mathbf{x}_N F_N = \frac{\partial F_R}{\partial t}$$

In the reduction of the remaining two terms we will make use of the fact that F_N must vanish at the limits of the integration. This is intuitively clear for the momentum integration, since the probability of finding a particle having infinite momentum is zero. For the integration in the physical space it is perhaps easiest to think in terms of extending the integration beyond the volume which is occupied by the system we are considering, so that F_N again vanishes at the limits of integration. For the case where the system occupies a definite volume defined by physical boundaries, that is, container walls, it is necessary to add another term to Equation (2–5) which contains the wall forces. However, it is easily shown that this term will not contribute to the equation for F_R if we are not interested in the behavior of F_R at the walls, which is generally the case, and so we will neglect this term here. We then have for the second term of the integrated Liouville equation

$$\begin{aligned}\int d\mathbf{x}_{R+1}\cdots d\mathbf{x}_N \sum_{i=1}^{N} \frac{\mathbf{p}_i}{m}\cdot\frac{\partial F_N}{\partial \mathbf{q}_i} &= \sum_{i=1}^{R} \frac{\mathbf{p}_i}{m}\cdot\frac{\partial}{\partial \mathbf{q}_i}\int d\mathbf{x}_{R+1}\cdots d\mathbf{x}_N F_N \\ &\quad + \int d\mathbf{x}_{R+1}\cdots d\mathbf{x}_N \sum_{i=R+1}^{N} \frac{\mathbf{p}_i}{m}\cdot\frac{\partial F_N}{\partial \mathbf{q}_i} \\ &= \sum_{i=1}^{R} \frac{\mathbf{p}_i}{m}\cdot\frac{\partial}{\partial \mathbf{q}_i}\int d\mathbf{x}_{R+1}\cdots d\mathbf{x}_N F_N \\ &= \sum_{i=1}^{R} \frac{\mathbf{p}_i}{m}\cdot\frac{\partial F_R}{\partial \mathbf{q}_i}\end{aligned}$$

Making use of Equation (2–5), the last term can be written as

$$\begin{aligned}-\int d\mathbf{x}_{R+1}\cdots d\mathbf{x}_N \sum_{i,j=1}^{N} \frac{\partial \phi_{ij}}{\partial \mathbf{q}_i}\cdot\frac{\partial F_N}{\partial \mathbf{p}_i} &= -\sum_{i,j=1}^{R} \frac{\partial \phi_{ij}}{\partial \mathbf{q}_i}\cdot\frac{\partial}{\partial \mathbf{p}_i}\int d\mathbf{x}_{R+1}\cdots d\mathbf{x}_N F_N \\ &\quad - \int d\mathbf{x}_{R+1}\cdots d\mathbf{x}_N \sum_{\substack{1\le i\le R\\ R+1\le j\le N}} \frac{\partial \phi_{ij}}{\partial \mathbf{q}_i}\cdot\frac{\partial F_N}{\partial \mathbf{p}_i} \\ &\quad - \int d\mathbf{x}_{R+1}\cdots d\mathbf{x}_N \sum_{\substack{R+1\le i\le N\\ 1\le j\le N}} \frac{\partial \phi_{ij}}{\partial \mathbf{q}_i}\cdot\frac{\partial F_N}{\partial \mathbf{p}_i}\end{aligned}$$

The last term on the right-hand side of the above equation vanishes because of the boundary conditions on F_N. Making use of the symmetry properties of F_N we can rewrite the second term as

$$-(N - R) \int d\mathbf{x}_{R+1} \sum_{i=1}^{R} \frac{\partial \phi_{iR+1}}{\partial \mathbf{q}_i} \cdot \frac{\partial F_{R+1}}{\partial \mathbf{p}_i} (\mathbf{x}_1, \mathbf{x}_2, \cdots, \mathbf{x}_{R+1}, t)$$

Finally, the first term is simply

$$\sum_{i,j=1} \frac{\partial \phi_{ij}}{\partial \mathbf{q}_i} \cdot \frac{\partial F_R}{\partial \mathbf{p}_i}$$

Combining the above results we get the following equation for F_R:

$$\frac{\partial F_R}{\partial t} + \sum_{i=1}^{R} \frac{\mathbf{p}_i}{m} \cdot \frac{\partial F_R}{\partial \mathbf{q}_i} - \sum_{i,j=1} \frac{\partial \phi_{ij}}{\partial \mathbf{q}_i} \cdot \frac{\partial F_R}{\partial \mathbf{p}_i}$$
$$= (N - R) \int d\mathbf{x}_{R+1} \sum_{i=1}^{R} \frac{\partial \phi_{iR+1}}{\partial \mathbf{q}_i} \cdot \frac{\partial F_{R+1}}{\partial \mathbf{p}_i} \qquad (2\text{–}6)$$

Equation (2–6) is called the BBGKY hierarchy of equations. The initials stand for the five people who originally and independently derived this equation: Bogoliubov, Born, H. S. Green, Kirkwood, and Yvon (the order is alphabetical and precisely antichronological). In particular, we will be interested in the first two hierarchy equations:

$$\frac{\partial F_1}{\partial t} + \frac{\mathbf{p}_1}{m} \cdot \frac{\partial F_1}{\partial \mathbf{q}_1} = (N - 1) \int d\mathbf{x}_2 \, \phi_{12}' \cdot \frac{\partial F_2}{\partial \mathbf{p}_1} \qquad (2\text{–}7)$$

$$\frac{\partial F_2}{\partial t} + \frac{\mathbf{p}_1}{m} \cdot \frac{\partial F_2}{\partial \mathbf{q}_1} + \frac{\mathbf{p}_2}{m} \cdot \frac{\partial F_2}{\partial \mathbf{q}_2} - \phi_{12}' \cdot \left(\frac{\partial}{\partial \mathbf{p}_1} - \frac{\partial}{\partial \mathbf{p}_2}\right) F_2$$
$$= (N - 2) \int d\mathbf{x}_3 \left\{ \phi_{13}' \cdot \left(\frac{\partial}{\partial \mathbf{p}_1} - \frac{\partial}{\partial \mathbf{p}_3}\right) + \phi_{23}' \cdot \left(\frac{\partial}{\partial \mathbf{p}_2} - \frac{\partial}{\partial \mathbf{p}_3}\right) \right\} F_3 \qquad (2\text{–}8)$$

Specification of proper initial data and computational facility do not represent the main problem in dealing with this equation, as is the case with the Liouville equation. Here the major problem is that Equation (2–6) is not closed; that is, the equation for F_R contains F_{R+1} as well, so that in the strict mathematical sense Equation (2–6) is not an equation. We are then faced with a different type of problem than that which is associated with the Liouville equation, namely the reduction of the coupled set of equations (2–6) to a determinate equation for F_R. In particular, we are interested in the closure of the F_1 equation, that is, removing the F_2 dependence in Equation (2–7) so that the equation becomes self contained in F_1.

2–2 REDUCTION OF THE FIRST BBGKY EQUATION TO THE BOLTZMANN EQUATION

The problem of obtaining a closed equation for F_1 from the BBGKY equations has been a problem of comparatively recent undertaking. Already in 1872 Boltzmann had obtained the (closed) equation for F_1 which bears his name; however, his method of deriving this equation was more or less *ad hoc* and not part of a systematic scheme. This shortcoming prevents any systematic generalization of his equation to situations which lay outside the regime of its applicability as restricted by the assumptions made in obtaining the equation. In obtaining an equation for F_1 from the Liouville equation (via the BBGKY equation) it becomes clear how such a generalization could be made. Of equal importance, the F_1 equation obtained in this way now acquires a pedigree, having a well-defined connection with the Liouville equation.

There have been many methods given for obtaining the Boltzmann equation from the BBGKY equation. For the purposes of this book the best suited is that given by Grad, which establishes very precisely the limit in which the equation is exact. As derived by Boltzmann, the F_1 equation is limited in its applicability to systems for which:

1. the density is sufficiently low so that only binary collisions between the constituent molecules need be considered;
2. the spatial dependence of gas properties is sufficiently slow so that collisions can be thought of as being localized in the physical space;
3. the interparticle potential is of sufficiently short range so that statement 1 is meaningful, for even in a rarefied gas the concept of binary collisions is meaningless if the potential is of such long range that a given particle interacts with many other particles at a particular instant.

Each of the above assumptions will also be included as part of the derivation of Boltzmann's equation given here. The first two follow from the limit which we consider, the third we introduce *ab initio*.

Prior to beginning the derivation of the Boltzmann equation, it is necessary to specify in mathematical terms the physical regime, or limit, which we will be describing. The number N of constituent molecules of the system we are describing is very large, on the order of 10^{23} molecules, and it is convenient to take the strict limit $N \to \infty$. (Note that the number of molecules in the system of interest, and the number of systems in the ensemble η, are different quantities.) At the same time the mass of each molecule is taken to approach zero in such a manner that the physically relevant quantity, the mass of the system mN, remains constant. We also take the size of the action sphere of a particle to approach zero in such a way that if σ is a parameter which characterizes the range of the interparticle forces, $N\sigma^2$ remains constant. The mean free path of the system can be shown to be proportional to $1/N\sigma^2$, so that in this limit the mean free path is constant. Thus in the sense that the limit we are considering describes a gas in which collisions take place (since the mean free path is finite), we are describing an imperfect gas, and the interparticle forces play a role in determining how F_1 changes in time. However, in the strict thermodynamic sense the description given is of a perfect gas; the thermodynamic properties retain their perfect gas values and there is no contribution to the transport of molecular properties from the interparticle forces. The reason for this is that in the limit we have described, the quantity $N\sigma^3$, the total action volume for the system goes to zero. The limit we are considering is then characterized by the following conditions:

$$\begin{aligned} N &\longrightarrow \infty \\ m &\longrightarrow 0 \\ \sigma &\longrightarrow 0 \\ N\sigma^2 &= \text{constant} \\ Nm &= \text{constant} \end{aligned} \tag{2-9}$$

The above limit is called the Boltzmann gas limit (BGL). It is only in this limit that many of the extremely technical properties of the Boltzmann equation of predominately mathematical interest (which we shall not be overly concerned with in this book) have been proved. Although we will only occasionally refer to this limit the reader should keep in mind that it is only in this limit, or an equivalent variant, that the classical theory of the Boltzmann equation is meaningful.

We now proceed with Grad's derivation of the Boltzmann equation. To this end we now consider the F_R to be functions of spatial coordinate and velocity instead of momentum, and accordingly consider phase points $\mathbf{y}_R = \mathbf{q}_R, \mathbf{v}_R$ instead of $\mathbf{x}_R$. Introducing truncated distributions $F_1{}^\sigma$, $F_2{}^\sigma$ where

$$F_1{}^\sigma \equiv \int_{D_1} d\mathbf{y}_2 \cdots d\mathbf{y}_N F_N \tag{2-10}$$

$$F_2{}^\sigma \equiv \int_{D_2} d\mathbf{y}_3 \cdots d\mathbf{y}_N F_N \tag{2-11}$$

with D_1 that part of the physical space such that no particle is within σ of particle 1, that is, $|\mathbf{q}_i - \mathbf{q}_1| \geq \sigma$, $i = 2, 3, \cdots, N$, and D_2 that part of the physical space such that $|\mathbf{q}_i - \mathbf{q}_1| \geq \sigma$, $i = 3, 4, \cdots, N$, it is clear that in the BGL we may replace F_1 by $F_1{}^\sigma$ and $F_2{}^\sigma$ by F_2. Integrating the Liouville equation over D_1 we get the BBGKY equation for $F_1{}^\sigma$:

$$\frac{\partial F_1{}^\sigma}{\partial t} + \mathbf{v}_1 \cdot \frac{\partial F_1{}^\sigma}{\partial \mathbf{q}_1} + (N-1) \oint_{S_2} d\mathbf{v}_2\, d\mathbf{S}_2 \cdot (\mathbf{v}_1 - \mathbf{v}_2) F_2{}^\sigma - (N-1) \frac{\partial}{\partial \mathbf{v}_1} \cdot \int_{|\mathbf{q}_1 - \mathbf{q}_2| > \sigma} d\mathbf{y}_2\, \phi_{12}{}' F_2{}^\sigma = 0 \quad (2\text{–}12)$$

where S_i is the surface of the sphere $|\mathbf{q}_i - \mathbf{q}_1| = \sigma$. Note that one more term appears in this equation than in the corresponding Equation (2–7) for F_1. This comes from the term

$$\sum_{i=1}^{N} \int_{D_1} d\mathbf{x}_2 \cdots d\mathbf{x}_N\, \mathbf{v}_i \cdot \frac{\partial F_N}{\partial \mathbf{q}_i} = \mathbf{v}_i \cdot \frac{\partial F_1{}^\sigma}{\partial \mathbf{q}_i}$$

$$+ (N-1) \int_{S_2} d\mathbf{v}_2\, d\mathbf{S}_2 \cdot \mathbf{v}_1 F_2{}^\sigma(\mathbf{y}_1,\mathbf{y}_2,t) - (N-1) \int_{S_2} d\mathbf{v}_2\, d\mathbf{S}_2 \cdot \mathbf{v}_2 F_2{}^\sigma(\mathbf{y}_1,\mathbf{y}_2,t)$$

where in carrying out the integration by parts we have had to take into account the fact that the limits of integration now depend on $\mathbf{q}_1$. For that part of the physical space in which $|\mathbf{q}_1 - \mathbf{q}_2| > \sigma$ we have $\phi_{12} = \phi_{12}{}' = 0$, so that the last term in Equation (2–12) is zero. The surface integral which appears in Equation (2–12) only contains contributions from particles completing a collision $[(\mathbf{v}_2 - \mathbf{v}_1) \cdot d\mathbf{S}_2 > 0]$ and particles initiating a collision $[(\mathbf{v}_2 - \mathbf{v}_1) \cdot d\mathbf{S}_2 < 0]$ and is thus different from the corresponding term which appears in the BBGKY equation, which has a continuous contribution from the term containing the effects of the interparticle forces. The former property is characteristic of the Boltzmann equation.

Introducing a coordinate system in the plane perpendicular to the vector $\mathbf{V} = \mathbf{v}_2 - \mathbf{v}_1$ with origin at $\mathbf{q}_1$ and defining the polar coordinate system r_1, ϵ in this plane (Figure 2–1) we see that each point on S_2 corresponds to a point on the disk $0 \leq r \leq \sigma$, $0 \leq \epsilon \leq 2\pi$. The disk is covered twice, once by projecting $S_2{}^+$, on which $\mathbf{V} \cdot d\mathbf{S}_2 > 0$, and once by projecting $S_2{}^-$, on which $\mathbf{V} \cdot d\mathbf{S}_2 < 0$. The points on $S_2{}^\pm$ are thus in a one-to-one correspondence with the points on the disk, and can thus be written as functions of r, and ϵ. Calling the element of area in the disk $d\omega = r\, dr\, d\epsilon$, and writing $\mathbf{q}^\pm = \mathbf{q}^\pm(r,\epsilon)$, we can transform the surface integration which appears in Equation (2–12) to an integration over the disk, giving

$$\frac{\partial F_1{}^\sigma}{\partial t} + \mathbf{v}_i \cdot \frac{\partial F_1{}^\sigma}{\partial \mathbf{q}_1} = (N-1) \int d\omega\, d\mathbf{v}_2\, V[F_2{}^\sigma(\mathbf{y}_1,\mathbf{y}_2{}^+,t) - F_2{}^\sigma(\mathbf{y}_1,\mathbf{y}_2{}^-,t)]$$

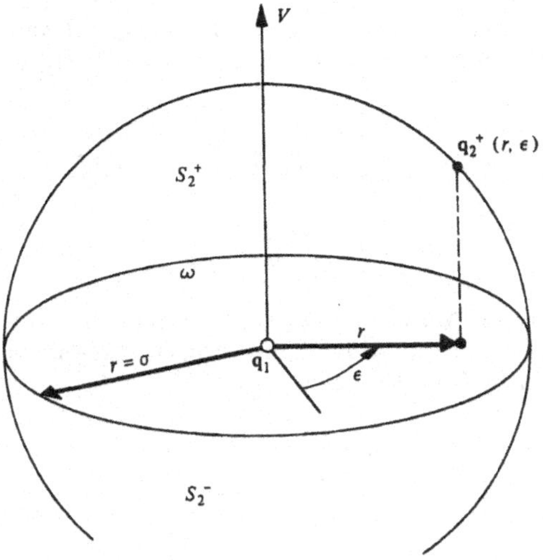

Figure 2–1

where $\mathbf{y}_2^\pm = \mathbf{q}_2^\pm, \mathbf{v}_2$, which as we have noted earlier we can write in the BGL as

$$\frac{\partial F_1}{\partial t} + \mathbf{v}_1 \cdot \frac{\partial F_1}{\partial \mathbf{q}_1} = N \int d\omega\, d\mathbf{v}_2\, V[F_2(\mathbf{y}_1,\mathbf{y}_2^+,t) - F_2(\mathbf{y}_1,\mathbf{y}_2^-,t)] \quad \textbf{(2–13)}$$

Note that the term on the right-hand side of Equation (2–13) is of order $N\sigma^2$ and thus remains finite.

In order to proceed from Equation (2–13) to the Boltzmann equation we now invoke the assumptions which we mentioned earlier. Accordingly we assume that the predominant collision process in the system is that of binary collisions. This implies that in Equation (2–8) we can neglect the integral term compared with the binary collision term when the latter is not zero, that is, when $|\mathbf{q}_1 - \mathbf{q}_2| < \sigma$, for times of order σ. For such times F_2 is described by a two-particle Liouville equation. The solution of the two-particle Liouville equation is simply F_2 = constant. We can therefore replace the term $F_2(\mathbf{y}_1,\mathbf{y}_2^+,t)$ which appears in Equation (2–13) by $F_2(\mathbf{Y}_1,\mathbf{Y}_2,\tau)$, where the $\mathbf{Y}_i$ are the phases which are transformed into the phases $\mathbf{y}_1$, $\mathbf{y}_2^+$ in a time $t - \tau$ by the motion described by the two-particle Liouville equation. We specifically choose the smallest possible value of τ for which the phase $\mathbf{Y}_1,\mathbf{Y}_2$ represents a pre-collisional phase which produces the post-collisional phase $\mathbf{y}_1$, $\mathbf{y}_2^+$. We now make the assumption that $F_2(\mathbf{y}_1,\mathbf{y}_2,t) = F_1(\mathbf{y}_1,t)F_1(\mathbf{y}_2,t)$ when $\mathbf{y}_1$, $\mathbf{y}_2$ are phases which represent molecules which have not yet collided. This assumption was also made by Boltzmann, and is referred to as the *stosszahlansatz* (collision number assumption). In the BGL it can be shown that this assumption is exact, since the error is of order σ. We will discuss this assumption in more detail in Chapter 5.

From the above discussion we see that the right-hand side of Equation (2–13) can be written as $N\int d\omega\, d\mathbf{v}_2 V[F_2(\mathbf{Y}_1,\mathbf{Y}_2,\tau) - F_2(\mathbf{y}_1,\mathbf{y}_2^-,t)]$ and since both of the phases which appear in the argument of F_2 are precollisional phases, we can correctly apply the *stosszahlansatz*. Finally, we note that in the BGL $\tau \to t$, and the physical space components of $\mathbf{Y}_1$, $\mathbf{Y}_2$ and $\mathbf{q}_2^- \to \mathbf{q}_1$, this replacement being strictly valid only if the spatial variation of F_1 is not appreciable over distances of order σ. This being the case we can then recast Equation (2–13) into a closed equation for F_1. Writing this now in terms of f, we have

$$\frac{\partial f}{\partial t} + \mathbf{v}_1 \cdot \frac{df}{\partial \mathbf{q}_1} = \frac{1}{m} \int d\omega\, d\mathbf{v}_2\, V[f(\mathbf{q}_1,\tilde{\mathbf{v}}_1,t)f(\mathbf{q}_1,\tilde{\mathbf{v}}_2,t) - f(\mathbf{q}_1,\mathbf{v}_1,t)f(\mathbf{q}_1,\mathbf{v}_2,t)] \quad (2\text{–}14)$$

where $\tilde{\mathbf{v}}_1$, $\tilde{\mathbf{v}}_2$ are the velocities which become $\mathbf{v}_1$, $\mathbf{v}_2$ following a binary collision.

Equation (2-14) is the celebrated Boltzmann equation. Some of its features are immediately recognizable. It is a nonlinear integral-differential equation. The left-hand side of the equation includes a differential operator in t and $\mathbf{q}$, and $\mathbf{v}$ appears as a parameter. The right-hand side is an integral operator in $\mathbf{v}$, and here t and $\mathbf{q}$ appear only as parameters. As a mathematical entity (perhaps beast would be more apt) Equation (2–14) is very complicated indeed, and in fact it was almost fifty years after its original derivation by Boltzmann that an approximate solution was obtained. However, even without solving this equation it is still possible to extract a good deal of useful information from it. In the next section we consider some of the ways in which this can be done.

2–3 THE MOMENTS OF f

The rationale for describing a many-body system in terms of f is that we can relate this description to the macroscopic description. In order to do this it is clear that we are going to have to somehow further contract the description as given by f, since even this description, minimally detailed on the molecular level, contains more information than the macroscopic description.

The mass density f has been defined such that $f\, d\mathbf{x}_1$ is the expected mass in the phase space volume element $d\mathbf{x}_1$ about $\mathbf{x}_1$. Integrating f over its velocity argument will then give an expression for the expected mass in the volume element $d\mathbf{q}_1$ about $\mathbf{q}_1$. Writing $\mathbf{q}$ for $\mathbf{q}_1$, we have

$$\rho(\mathbf{q},t) = \int d\mathbf{v}_1\, f(\mathbf{q},\mathbf{v}_1,t) \tag{2–15}$$

The function ρ is the macroscopic fluid mass density which is used in the fluid mechanical description.

Defining the moments of f so that the νth moment is given by

$$M_{\alpha\beta\cdots\nu} = \int d\mathbf{v}_1\, \mathbf{v}_{1\alpha}\mathbf{v}_{1\beta} \cdots \mathbf{v}_{1\nu} f(\mathbf{q},\mathbf{v}_1,t) \tag{2–16}$$

we see that ρ is the zeroth moment of f. The first moment also has a physical significance; we have

$$\rho(\mathbf{q},t)\mathbf{u}(\mathbf{q},t) = \int d\mathbf{v}_1\, \mathbf{v}_1 f(\mathbf{q},\mathbf{v}_1,t) \tag{2–17}$$

The quantity $\mathbf{u}$ is the macroscopic fluid flow velocity.

It is now convenient to introduce the quantity $\mathbf{v}_0 = \mathbf{v}_1 - \mathbf{u}$, the so-called peculiar velocity, which is the particle velocity with respect to the macroscopic fluid flow velocity. This quantity can be used to define the temperature of the system by making use of the equipartition theorem, a well-known result of equilibrium statistical mechanics. This theorem states that for any system the internal energy per degree of freedom is $kT/2$. Therefore, for a system of structureless molecules such as we are considering, the internal energy per unit mass is $\frac{3}{2}RT$. The total energy in the BGL is due to kinetic energy only, so that the energy density per unit mass is

$$\frac{1}{2\rho}\int d\mathbf{v}_1\, v_1^2 f = \frac{1}{2\rho}\int d\mathbf{v}_1\, v_0^2 f + \tfrac{1}{2}u^2$$

The first term is the contribution of the particle motion to the energy of the system. This is the fluid internal energy, which we will denote by ξ, and which, from the equipartition theorem, must satisfy the condition

$$\xi = \tfrac{3}{2}RT \tag{2–18}$$

As we will see below this result is compatible with the equation of state for the system, which in the BGL is given by the perfect gas law.

To complete the identification of the macroscopic thermo-fluid variables with the moments of f we must find relationships expressing the pressure tensor and the heat flux vector in terms of these moments. In the usual treatments of the macroscopic theory these former quantities are not treated as independent variables. However, as we shall shortly see, one of the advantages of the microscopic description is that these variables can be so treated, that is, the microscopic description is completely self contained.

Let us first consider the pressure tensor. We first consider a system in which there is no macroscopic fluid motion, so that $\mathbf{v}_0 = \mathbf{v}_1$. For this system the pressure tensor $P_{\alpha\beta}$, is defined in the following way. If we imagine a surface $d\mathbf{S}$ arbitrarily placed into the system, then the force acting over this imaginary surface is $\mathbf{P} \cdot d\mathbf{S}$. Each molecule in the system carries with it an amount of momentum $m\mathbf{v}$, and since the rate at which molecules flow through $d\mathbf{S}$ is $F_1\mathbf{v}_1 \cdot d\mathbf{S}$, the net rate at which momentum flows through $d\mathbf{S}$ is then $d\mathbf{S} \cdot \int d\mathbf{v}_1\, \mathbf{v}_1\mathbf{v}_1 f$. We can therefore identify

$$\mathbf{P} = \int d\mathbf{v}_1\, \mathbf{v}_1\mathbf{v}_1 f$$

The hydrostatic pressure p acting on $d\mathbf{S}$ is the force normal to this surface divided by $d\mathbf{S}$, so that, for example, if $d\mathbf{S}$ is normal to the x direction

$$p = \int d\mathbf{v}_1\, v_{1x}^2 f$$

In this case the components of $\mathbf{P}$ in the y-z plane represent the shearing forces acting on the surface.

For a gas which has a net macroscopic fluid flow velocity $\mathbf{u}$, the pressure tensor is defined in the same way as above, with the exception that now the imaginary surface $d\mathbf{S}$ moves with the fluid velocity $\mathbf{u}$. We then have the completely general result

$$\mathbf{P} = \int d\mathbf{v}_1\, \mathbf{v}_0\mathbf{v}_0 f \tag{2–19}$$

Note that $\mathbf{P}$ is symmetric, that is, $P_{\alpha\beta} = P_{\beta\alpha}$. We will see later that the diagonal components of $\mathbf{P}$ are equal, so that

$$p = \frac{1}{3}\int d\mathbf{v}_1\, \mathbf{v}_0^2 f \tag{2–20}$$

From the definition of ξ we then have

$$p = \tfrac{2}{3}\rho\xi = \rho RT \tag{2–21}$$

Thus we recover the well-known equation of state for a perfect gas, which serves as a check for the consistency of our identification of molecular quantities with macroscopic properties.

In considering the transport of energy through the fluid we must be careful to distinguish between the net energy flow $\mathbf{E}$ and the heat flux $\mathbf{Q}$. The former quantity is the total kinetic energy flow per unit area through an imaginary fixed surface, whereas the latter quantity is the total kinetic energy flow per unit area through an imaginary surface which moves with the macroscopic fluid flow velocity. The heat flux is thus only a component of the energy flux; it is the energy flux through the fluid. Accordingly we have

$$\mathbf{Q} = \int d\mathbf{v}_1\, \frac{v_0^2}{2}\, \mathbf{v}_0 f \tag{2–22}$$

and

$$\begin{aligned}\mathbf{E} &= \int d\mathbf{v}_1 \frac{v_1^2}{2} \mathbf{v}_1 f \\ &= \int d\mathbf{v}_1 \tfrac{1}{2}[v_0^2 + 2\mathbf{u}\cdot\mathbf{v}_0 + u^2][\mathbf{v}_0 + \mathbf{u}]f \\ &= \mathbf{Q} + \mathbf{u}\cdot\mathsf{P} + \rho\mathbf{u}(\xi + \tfrac{1}{2}u^2)\end{aligned} \tag{2-23}$$

The total energy flow is thus seen to be made up of heat, work, and convected energy.

In deriving the above results we have neglected any transport of energy or momentum at a distaace through the mechanism of the intermolecular force field. This is consistent with the BGL in which these effects give no contribution. It is possible to include these effects as well; for example to consider an alternate limit. However, the precise way in which this can be done is still an open question and so we will not consider this problem here.

2-4 THE HYDRODYNAMICAL EQUATIONS

Just as we were able to obtain an equation for the F_R by integrating Liouville's equation, so we can also obtain equations for the $M_{\alpha\beta\cdots\nu}$ by multiplying the Boltzmann equation by 1, $\mathbf{v}_1$, $\mathbf{v}_1\mathbf{v}_1$, $\cdots$ and integrating over $\mathbf{v}_1$. Before doing this it will be necessary to quote a result which we shall prove in the next chapter. 1, $\mathbf{v}_1$, v_1^2 are eigenfunctions of the integral operator $\int d\mathbf{v}_1\, J(f)$ having eigenvalue zero. This means that

$$\int d\mathbf{v}_1\, J(f) \begin{bmatrix} 1 \\ \mathbf{v}_1 \\ v_1^2 \end{bmatrix} = 0 \tag{2-24}$$

This property has a simple physical interpretation which we will also discuss in the next chapter.

Multiplying the Boltzmann equation successively by 1, $\mathbf{v}_1$, $\frac{1}{2}v_1^2$ and integrating over $\mathbf{v}_1$, gives, after some rearranging of terms, the following equations (it is suggested that the reader do this himself):

$$\frac{\partial\rho}{\partial t} + \frac{\partial}{\partial\mathbf{q}}(\rho\mathbf{u}) = 0 \tag{2-25}$$

$$\frac{\partial}{\partial t}(\rho u_\alpha) + \frac{\partial}{\partial q_\beta}(\rho u_\alpha u_\beta + P_{\beta\alpha}) = 0 \tag{2-26}$$

$$\frac{\partial}{\partial t}[\rho(\xi + \tfrac{1}{2}u^2)] + \frac{\partial}{\partial q_\beta}[\rho u_\beta(\xi + \tfrac{1}{2}u^2)] + u_\alpha P_{\alpha\beta} + Q_\beta = 0 \tag{2-27}$$

The above equations are the usual equations of the macroscopic hydrodynamical theory for mass, momentum, and energy conservation. At this point we see vividly illustrated one of the major advantages of the microscopic description. The macroscopic conservation equations considered by themselves constitute an open set of equations. We have but five equations to describe 13 independent variables. Therefore, the hydrodynamical description is not self-contained. Making different assumptions to close these equations leads to the various macroscopic descriptions, that is, the Euhler equations and Navier–Stokes equations; from the microscopic point of view these descriptions are only approximations, and never exact. The fact is that in the microscopic description the macroscopic conservation equations are not needed to find the space and time dependence of the macroscopic variables; in fact we do not need to use these equations at all. Instead, this information is found directly and exactly from a knowledge of f, which is obtained by solving the Boltzmann equation.

We consider very briefly two other advantages of the microscopic description. There are specific instances where the hydrodynamical equations are correct in principle, but incorrect in detail (for example, for describing a weak shock wave). There are also instances where the hydrodynamical equations are incorrect in principle, for example, when the system of interest is so rarefied that collisions of any kind are a rarity. For such a system, called a Knudsen gas, a hydrodynamical description is *a priori* precluded since we are no longer dealing with a continuum. This case is clearly not included in the BGL, and in order to describe such a system we must consider a new limit called the Knudsen limit. In this limit the condition $N\sigma^2 \to$ constant which appears in the BGL is replaced by the condition $N\sigma^2 \to 0$, so that the mean free path becomes infinite.

There are other cases in addition to the Knudsen gas for which the microscopic description in the BGL limit is incorrect, for example, a dense gas. Even here, however, the microscopic description, in principle, is to be preferred to the macroscopic description. However, for this case the correct formulation of the microscopic description becomes, as we have seen, a formidable problem and by virtue of its simplicity the macroscopic description is at present the most widely used.

References

The derivation of Boltzmann's equation given here follows that given by Grad (1958). For other derivations, which treat Boltzmann's equation as an approximation rather than an exact contraction of Liouville's equation, see:

1. N. N. Bogoliubov, in *Studies in Statistical Mechanics*, vol. 1, J. deBoer and G. E. Uhlenbeck, Eds. New York: Wiley, 1962.

2. M. Born and H. S. Green, *A General Kinetic Theory of Liquids.* Cambridge, England: Cambridge University Press, 1949.

3. E. G. D. Cohen, in *Lectures in Theoretical Physics*, vol. 9C, W. E. Brittin, A. Barut, and M. Guenin, Eds. New York: Gordon and Breach, 1967.

4. M. S. Green, *J. Chem. Phys.*, vol. 25, p. 836, 1956.

5. J. G. Kirkwood, *J. Chem. Phys.*, vol. 15, p. 72, 1947.

6. I. Prigogine, *Non-Equilibrium Statistical Mechanics.* New York: Wiley, 1962.

7. J. Yvon, *La Theorie Statistique des Fluides et l'Equation d'Etat.* Paris: Hermann, 1935.

Problems

2–1. Derive the first BBGKY equation directly from the second BBGKY equation.

2–2. Derive the BBGKY equations for a system which is acted on by a time independent external field.

2–3. Derive the BBGKY equations for the case where the forces exerted on the system by its container are taken into account by adding a term

$$\sum_{i=1}^{N} U(\mathbf{q}_i)$$

to the Hamiltonian. The quantity $U(\mathbf{q}_i)$, which is the potential energy of the ith particle due to the container, is zero for $\mathbf{q}_i$ inside the container, and infinite for $\mathbf{q}_i$ outside the container.

2–4. Obtain the equation for $F_2(\mathbf{q}_1,\mathbf{q}_2,t)$ by taking the appropriate moments of the second BBGKY equation.

2–5. Obtain the equation for $F_2(\mathbf{p}_1,\mathbf{p}_2,t)$ by taking the appropriate moments of the second BBGKY equation.

2–6. Derive the conservation equations for mass and momentum directly from the first BBGKY equation. Show that the momentum equation can be written in a form identical to Equation (2–26) if P is suitably redefined, and show that this new P reduces to the expression given by Equation (2–19) when the BGL is taken.

2–7. With respect to the results of Problem 2–6, why is the mass conservation equation which is derived from the BBGKY equation the same as that derived from Boltzmann's equation, and why do the momentum conservation equations which are found from these two equations differ?

2–8. An equation of change for the temperature can be derived either directly from Boltzmann's equation by taking the appropriate moments, or directly from Equations (2–25) to (2–27). Verify the following form of this equation by both methods:

$$\frac{\partial T}{\partial t} + \mathbf{u} \cdot \frac{\partial T}{\partial \mathbf{q}} + \frac{2}{3R\rho}\left(\mathsf{P} \cdot \frac{\partial \mathbf{u}}{\partial \mathbf{q}}\right) + \frac{\partial}{\partial \mathbf{q}} \cdot \mathbf{Q} = 0$$

2–9. Derive the macroscopic conservation Equations (2–25) to (2–27) by taking the appropriate moments of Boltzmann's equation.

2–10. Show that if

$$f = \frac{\rho}{(2\pi RT)^{3/2}} e^{-(\mathbf{v}-\mathbf{u})^2/2RT}$$

then

$$J(f) = 0$$

Prove that if f is given by the above expression and is also independent of $\mathbf{q}$, then ρ, $\mathbf{u}$, T must be independent of both $\mathbf{q}$ and t.

CHAPTER 3

description of binary collisions

3-1 GENERAL FORMULATION

It is clear that in order to entertain solutions to Boltzmann's equation we are going to have to prescribe the explicit dependence of the dependent variables $\bar{\mathbf{v}}_1$, $\bar{\mathbf{v}}_2$ (which appear in the argument of two of the f's in $J(f)$) on the independent variables, which we have specified as $\mathbf{v}_1$, $\mathbf{v}_2$, r, and ϵ. In order to do this we will have to consider in some detail the properties of classical binary collisions. The information which we require concerning this collision process is actually quite minimal; we need no information about the particle orbits themselves, and we need not even specify the instantaneous particle velocities along these trajectories. The information that we do require pertains only to the asymptotic collisional velocity states (both pre-collisional and post-collisional). It is important to take notice of the fact that we have assumed that collisions are discontinuous events, in the sense that we can specify states before and after the event. For other than hard sphere molecules (see Section 3-3) the forces between the molecules are of an infinite range, although beyond a certain separation these forces are very weak. In practice we will have to exercise some criteria for cutting off the range of these forces. The reason for this, aside from the fact that it allows us to unambiguously define asymptotic collisional states, is that the Boltzmann collision integral,[1]

$$J(f) = \frac{1}{m} \int d\omega\, d\mathbf{v}_2\, V[\bar{f}_1\bar{f}_2 - f_1 f_2] \tag{3-1}$$

is not absolutely convergent since $\int d\omega$ is infinite. Therefore we must rely on cancellations in the collision integral to produce a convergent

[1] We adopt the following notation for future use:

$$\bar{f}_i \equiv f(\mathbf{q},\bar{\mathbf{v}}_i,t), \qquad f_i \equiv f(\mathbf{q},\mathbf{v}_i,t), \qquad f_i' \equiv f(\mathbf{q},\mathbf{v}_i',t)$$

result. It is preferable however to effect convergence directly through a cutoff of the intermolecular forces.

Let us first consider the analytical relationship connecting the pre-collisional velocities $\mathbf{v}_1$, $\mathbf{v}_2$ to their post-collisional values, which we now define as $\mathbf{v}_1'$, $\mathbf{v}_2'$. In mathematical terms, we seek the transformation T which takes $\mathbf{v}_1$, $\mathbf{v}_2$ into $\mathbf{v}_1'$, $\mathbf{v}_2'$. We will then use this relationship to determine the connection between $\bar{\mathbf{v}}_1$, $\bar{\mathbf{v}}_2$, the pre-collisional state which determines the post-collisional state $\mathbf{v}_1$, $\mathbf{v}_2$. Note that it is the variables $\bar{\mathbf{v}}_1$, $\bar{\mathbf{v}}_2$ which explicitly appear in $J(f)$ and not $\mathbf{v}_1'$, $\mathbf{v}_2'$.

In a classical binary collision both linear momentum and energy are conserved quantities. Since in the asymptotic collisional states the particles are not exerting forces on each other, the total energy of each particle in such states is kinetic and the conservation equations relating the asymptotic collisional states are

$$\mathbf{v}_1 + \mathbf{v}_2 = \mathbf{v}_1' + \mathbf{v}_2' \tag{3–2}$$
$$v_1^2 + v_2^2 = v_1'^2 + v_2'^2 \tag{3–3}$$

Since we have six scalar unknowns $(\mathbf{v}_1', \mathbf{v}_2')$ and only four scalar equations, Equations (3–2) and (3–3), we will have to specify two parameters in addition to $\mathbf{v}_1$, $\mathbf{v}_2$ to describe the collision. This is conveniently done by noting that the above equations have the following geometrical interpretation. If we construct the polygon (Figure 3–1) having as diagonals the vectors $\mathbf{V} = \mathbf{v}_2 - \mathbf{v}_1$ and $\mathbf{V}' = \mathbf{v}_2' - \mathbf{v}_1'$, this polygon is in fact a rectangle.

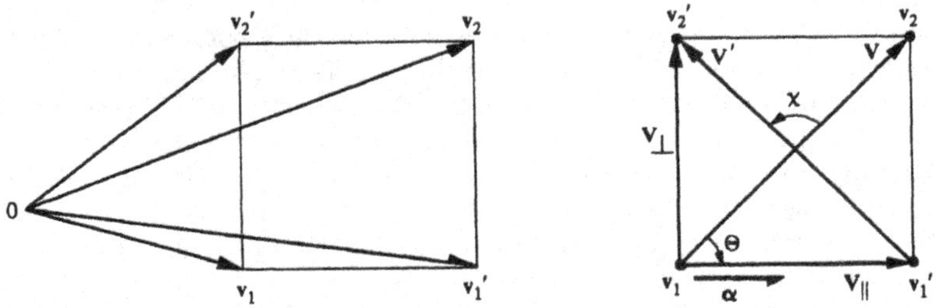

Figure 3–1 Velocities in a classical binary collision.

That this follows from Equations (3–2) and (3–3) can be seen from the following expressions which may be derived from these equations:

$$\mathbf{v}_1' - \mathbf{v}_1 = \mathbf{v}_2 - \mathbf{v}_2' \tag{3–4}$$
$$|\mathbf{v}_2 - \mathbf{v}_1| = |\mathbf{v}_2' - \mathbf{v}_1'| \tag{3–5}$$

Equation (3–4) indicates that the polygon is a parallelogram, and Equation (3–5) indicates that the parallelogram is a rectangle (this equation also tells us that the magnitude of the relative velocity, $|\mathbf{V}|$, is not changed by the collision). The additional two parameters which we require to complete the description of the collision can now be embodied in the unit vector $\boldsymbol{\alpha}$ directed along the vector $\mathbf{v}_1' - \mathbf{v}_1$. We will see later that $\boldsymbol{\alpha}$ can be related to the variables r, ϵ which appear in the collision integral.

With $\boldsymbol{\alpha}$ as defined above, we now introduce the scalar quantity a through the relationship

$$\mathbf{v}_1' - \mathbf{v}_1 = a\boldsymbol{\alpha} \tag{3-6}$$

so that (see Figure 3-1)

$$\mathbf{v}_1' = \mathbf{v}_1 + a\boldsymbol{\alpha} \tag{3-7}$$

$$\mathbf{v}_2' = \mathbf{v}_2 - a\boldsymbol{\alpha} \tag{3-8}$$

Substituting these equations into Equation (3-3) we get the following relationship:

$$a\boldsymbol{\alpha} \cdot (\mathbf{v}_2 - \mathbf{v}_1) = a^2 \tag{3-9}$$

and if $a \neq 0$ (see below)

$$a = \boldsymbol{\alpha} \cdot \mathbf{V} \tag{3-10}$$

so that

$$\mathbf{v}_1' = \mathbf{v}_1 + \boldsymbol{\alpha}(\boldsymbol{\alpha} \cdot \mathbf{V}) \tag{3-11}$$

$$\mathbf{v}_2' = \mathbf{v}_2 - \boldsymbol{\alpha}(\boldsymbol{\alpha} \cdot \mathbf{V}) \tag{3-12}$$

This is the result we have been looking for; Equations (3-11) and (3-12) express $\mathbf{v}_1', \mathbf{v}_2'$ in terms of $\mathbf{v}_1, \mathbf{v}_2$ and the collision parameter $\boldsymbol{\alpha}$ (the quantity $\boldsymbol{\alpha}$ is sometimes referred to as the apse direction, a term used in rigid body dynamics). Note that the special case $(\boldsymbol{\alpha} \cdot \mathbf{V}) = 0$ corresponds to the identity transformation $\mathbf{v}_1, \mathbf{v}_2 \rightarrow \mathbf{v}_1, \mathbf{v}_2$.

Equations (3-11) and (3-12) define the transformation T that takes $\mathbf{v}_1, \mathbf{v}_2$ into $\mathbf{v}_1', \mathbf{v}_2'$. Let us now consider the inverse of this transformation, T^{-1}, which takes $\mathbf{v}_1', \mathbf{v}_2'$ into $\mathbf{v}_1, \mathbf{v}_2$. From the equations which define T we have

$$\mathbf{V}' = \mathbf{V} - 2\boldsymbol{\alpha}(\boldsymbol{\alpha} \cdot \mathbf{V}) \tag{3-13}$$

so that, taking the scalar product of each side of this equation with $\boldsymbol{\alpha}$, we obtain the following relationship:

$$\boldsymbol{\alpha} \cdot \mathbf{V}' = -\boldsymbol{\alpha} \cdot \mathbf{V} \tag{3-14}$$

which with T serves to define T^{-1}:

$$\mathbf{v}_1 = \mathbf{v}_1' + \boldsymbol{\alpha}(\boldsymbol{\alpha} \cdot \mathbf{V}') \tag{3-15}$$

$$\mathbf{v}_2 = \mathbf{v}_2' - \boldsymbol{\alpha}(\boldsymbol{\alpha} \cdot \mathbf{V}') \tag{3-16}$$

Comparing Equations (3-11) and (3-12) with Equations (3-15) and (3-16), we see that the transform T is involutive for fixed $\boldsymbol{\alpha}$ (or with $\boldsymbol{\alpha}$ reversed in sign), that is,

$$T = T^{-1} \tag{3-17}$$

As a result of Equation (3–17) the Jacobian of both the transformation T and T^{-1} is unity, so that we have, for example,

$$d\mathbf{v}_1\, d\mathbf{v}_2 = |J|\, d\mathbf{v}_1'\, d\mathbf{v}_2' = d\mathbf{v}_1'\, d\mathbf{v}_2' \tag{3–18}$$

where J is defined in the standard way:

$$J = \frac{\partial(\mathbf{v}_1', \mathbf{v}_2')}{\partial(\mathbf{v}_1, \mathbf{v}_2)} \equiv \begin{vmatrix} \frac{\partial v_{1_x}'}{\partial v_{1_x}} & \frac{\partial v_{1_x}'}{\partial v_{1_y}} & \cdots & \frac{\partial v_{1_x}'}{\partial v_{2_z}} \\ \frac{\partial v_{1_y}'}{\partial v_{1_x}} & & & \\ \vdots & & & \vdots \\ \frac{\partial v_{2_z}'}{\partial v_{1_x}} & & \cdots & \frac{\partial v_{2_z}'}{\partial v_{2_z}} \end{vmatrix} \tag{3–19}$$

The above result may be verified by readers unfamiliar with the properties of involutive transformations by directly computing the determinant. Repeated use will be made of both equations in the development of the properties of the collision term in the following chapter.

As we mentioned earlier, the velocity state which appears in the Boltzmann equation is $\bar{\mathbf{v}}_1$, $\bar{\mathbf{v}}_2$, the velocities which go into $\mathbf{v}_1$, $\mathbf{v}_2$. From Equations (3–7), (3–8), (3–11), and (3–12), we see that $\bar{\mathbf{v}}_1$, $\bar{\mathbf{v}}_2$ can be replaced by $\mathbf{v}_1'$, $\mathbf{v}_2'$ in the Boltzmann equation providing that the collision parameter $\boldsymbol{\alpha}$ is reversed (see Figure 3–2).

It still remains for us to explicitly determine the relationship between the parameter $\boldsymbol{\alpha}$ and the integration variable $d\omega$ which appears in the collision term. To this end consider a spherical coordinate system with origin fixed on molecule 1, and with polar (z) axis defined by the relative velocity $\mathbf{V}$ (Figure 3–3). The azimuthal angle Θ is defined by the equation

$$\boldsymbol{\alpha} \cdot \mathbf{V} = V \cos \Theta \tag{3–20}$$

From Figure 3–1 we see that the angle Θ is a measure of the angle through which the relative velocity is turned in the collision; specifically, the angle of deflection is $\pi - 2\Theta$. The angle in the plane perpendicular to $\mathbf{V}$, measured with respect to some arbitrary origin, is ϵ; further, r is the intercept in this plane of the post-collisional asymptote of the trajectory of particle 2. The quantity r is called the impact parameter. For purely repulsive intermolecular potentials r is also the distance of closest approach. For a given intermolecular potential we will have $r = r\,(\Theta,V)$

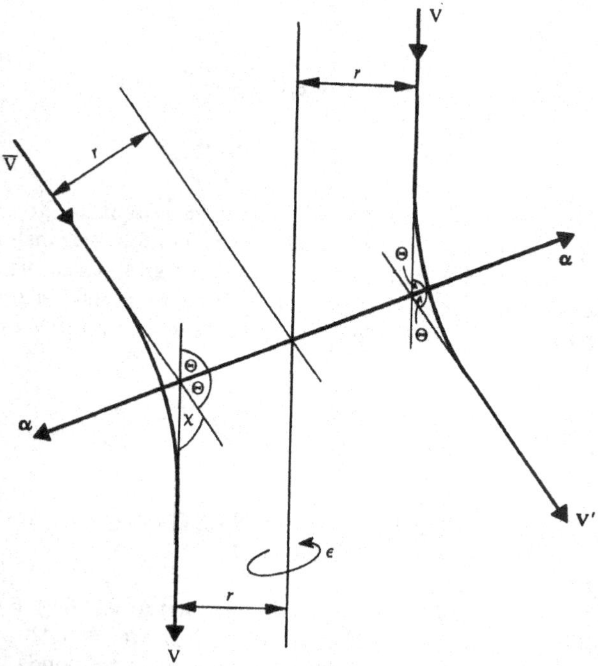

Figure 3–2 Direct collision which appears in Boltzmann equation as derived from BBGKY equation in BGL and inverse collision which we use to replace direct collision in Boltzmann equation.

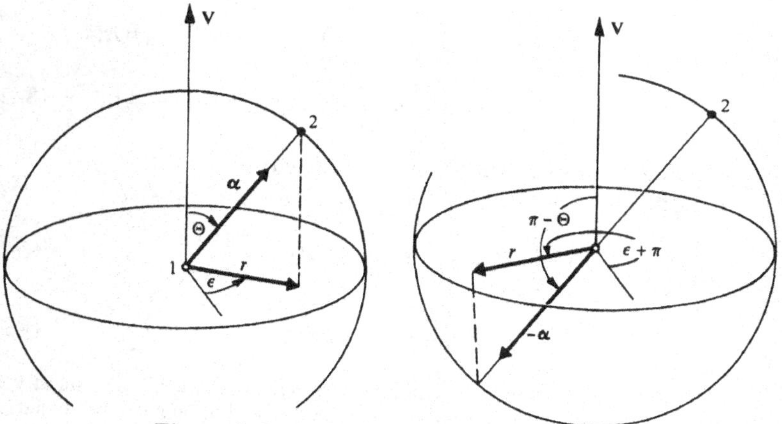

Figure 3–3 The transformation $\boldsymbol{\alpha} \rightarrow -\boldsymbol{\alpha}$.

specified, so that

$$d\omega = r\,dr\,d\epsilon = r(\theta,V)\left|\frac{\partial r(\theta,V)}{\partial\theta}\right| d\theta\,d\epsilon$$
$$\equiv \frac{1}{V}B(\theta,V)\,d\theta\,d\epsilon \tag{3-21}$$

We will restrict ourselves to purely repulsive potentials here, in which case Θ is a single-valued function of r and the above transformation is valid. For potentials which are both repulsive and attractive $r|\partial r/\partial\Theta|$ is a multivalued function of θ, and $B(\Theta,V)$ must be redefined by including all its values at given Θ. The Boltzmann collision term can now be written as

$$J(f) = \frac{1}{m}\int d\mathbf{v}_2\,d\Theta\,d\epsilon\,B(\Theta,V)[f_1'f_2' - f_1f_2] \tag{3-22}$$

3-2 AN ALTERNATE REPRESENTATION OF THE COLLISION INTEGRAL

For purely repulsive intermolecular potentials $0 \leq \theta \leq \pi/2$, and the angular integrations which appear in $J(f)$ are over the hemisphere $0 \leq \theta \leq \pi/2$, $0 \leq \epsilon \leq 2\pi$. Setting $\boldsymbol{\alpha} \to -\boldsymbol{\alpha}$ corresponds to setting $\Theta \to \pi - \theta$, $\epsilon \to \pi + \epsilon$ (see Figure 3-3) so that for repulsive potentials we can define $B(\Theta,V) = B(\pi - \theta, V)$. We can then extend the angular integration appearing in $J(f)$ over the entire sphere, $0 \leq \theta \leq \pi$, $0 \leq \epsilon \leq 2\pi$, providing we put a factor $\frac{1}{2}$ in front of the integral, that is,

$$J(f) = \frac{1}{m}\int_0^{\pi/2} d\theta \int_0^{2\pi} d\epsilon \int d\mathbf{V}\,B(\theta,V)[f_1'f_2' - f_1f_2]$$
$$= \frac{1}{2m}\int_0^{\pi} d\theta \int_0^{2\pi} d\epsilon \int d\mathbf{V}\,B(\theta,V)[f_1'f_2' - f_1f_2] \tag{3-23}$$

where we have also written $d\mathbf{V} = d\mathbf{v}_2$, $\mathbf{v}_1$ fixed.

Introducing now the function

$$G = G(|V\cos\theta|,|V\sin\theta|) \equiv \frac{B(\theta,V)}{|\sin\theta|} \tag{3-24}$$

and noting that[2]

$$d_2\alpha = \sin\theta\,d\theta\,d\epsilon \tag{3-25}$$

[2] In this section only, to avoid possible confusion, we will use the subscript 2 to refer to both differential in the plane, as in $d_2\omega$, and differential solid angles, as in $d_2\alpha$ about $\boldsymbol{\alpha}$.

we have the following relationship:

$$G(|V\cos\theta|,|V\sin\theta|)d_2\alpha = B(\theta,V)\,d\theta\,d\epsilon \tag{3–26}$$

We can therefore write

$$J(f) = \frac{1}{2m}\oint d_2\alpha\,d\mathbf{V}\,G(|V\cos\theta|,|V\sin\theta|)[f_1'f_2' - f_1f_2] \tag{3–27}$$

where the $\oint$ indicates that $d_2\alpha$ is to be integrated over the entire unit sphere. As mentioned earlier, we assume that some suitable cutoff has been applied to the intermolecular potential, so that

$$\int d_2\omega,\ \int d\theta\,B(\theta,V),\ \oint d_2\alpha G(|V\cos\theta|,|V\sin\theta|) < \infty \tag{3–28}$$

Therefore it follows that the quantity

$$\nu(\mathbf{v}_1) \equiv \frac{1}{m}\int d_2\omega\,d\mathbf{V}\,Vf_2 < \infty \tag{3–29}$$

exists, and that we can write

$$J(f) = \frac{1}{2m}\oint d_2\alpha\,d\mathbf{V}\,G(|V\cos\theta|,|V\sin\theta|)f_1'f_2' - \nu(\mathbf{v}_1)f_1 \tag{3–30}$$

The vector $\mathbf{V}$ can be decomposed into a component parallel to $\boldsymbol{\alpha}$, $V_\parallel\boldsymbol{\alpha}$ and a component perpendicular to $\boldsymbol{\alpha}$, $\mathbf{V}_\perp$. We can therefore set $d\mathbf{V} = dV_\parallel\,d_2V_\perp$ (so that $V_\parallel$ is integrated along a line and $\mathbf{V}_\perp$ over a plane). Since $\mathbf{V}_\parallel = V_\parallel\boldsymbol{\alpha}$, we have

$$d\mathbf{V}_\parallel = \tfrac{1}{2}V_\parallel^2\,dV_\parallel\,d_2\alpha$$

$$\frac{2}{V_\parallel^2}d\mathbf{V}_\parallel\,d_2V_\perp = dV_\parallel\,d_2V_\perp d_2\alpha = d_2\alpha\,d\mathbf{V} \tag{3–31}$$

and we can replace the differential $d_2\alpha\,d\mathbf{V}$ which appears in $J(f)$, which can then be written as

$$\frac{1}{m}\int d_2V_\perp\,d\mathbf{V}_\parallel\frac{G(V_\parallel,V_\perp)}{V_\parallel^2}f_1'f_2' - \nu(\mathbf{v}_1)f_1$$

The velocity arguments of the f's in the above expression can now be explicitly expressed in terms of the integration variables since (see Figure 3–1)

$$\mathbf{v}_1' = \mathbf{v}_1 + \mathbf{V}_\parallel$$
$$\mathbf{v}_2' = \mathbf{v}_1 + \mathbf{V}_\perp \tag{3–32}$$

so that we have, finally

$$J(f) = \frac{1}{m}\int d_2V_\perp\,d\mathbf{V}_\parallel\frac{G(V_\parallel,V_\perp)}{V_\parallel^2}f(\mathbf{v}_1+\mathbf{V}_\parallel)f(\mathbf{v}_1+\mathbf{V}_\perp) - \nu(\mathbf{v}_1)f_1 \tag{3–33}$$

A still simpler form for the collision integral will be given in the next chapter for the case where f is only slightly perturbed from its equilibrium value, so that $J(f)$ can be linearized.

3–3 HARD SPHERE MOLECULES

We shall now digress briefly from the development of the general theory to present a specific result—the evaluation of the quantity $B(\Theta,V)$ for hard sphere molecules. The hard sphere molecule, also referred to as the billard ball model, is often used as a model for molecules which interact through a potential that is short-ranged and strongly repulsive. This model qualitatively reproduces the behavior of such molecules and has the added virtue of being generally quite simple to treat analytically. As we shall see, in the theory of the Boltzmann equation this is in fact not the case, and it is rather the so-called Maxwell molecule model (described in Section 3–5) which entails the fewest computational difficulties. For the specific purpose of calculating $B(\Theta,V)$, however, a hard sphere model does offer the least analytical difficulty. In this model the molecules are considered to be rigid impenetrable spheres of radius r_0. The intermolecular potential consequently has the behavior shown in Figure 3–4.

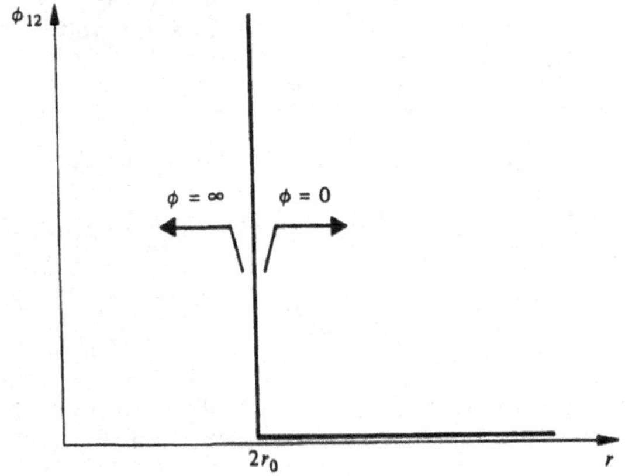

Figure 3–4 Hard sphere potential.

As shown in Figure 3–3, $\boldsymbol{\alpha}$ will be in the direction of the line connecting the centers of the two colliding molecules at the instant that the collision is initiated, so that we will have the configuration shown in Figure 3–5. Note that in this special case $\boldsymbol{\alpha}$ will be independent of $\mathbf{V}$. The

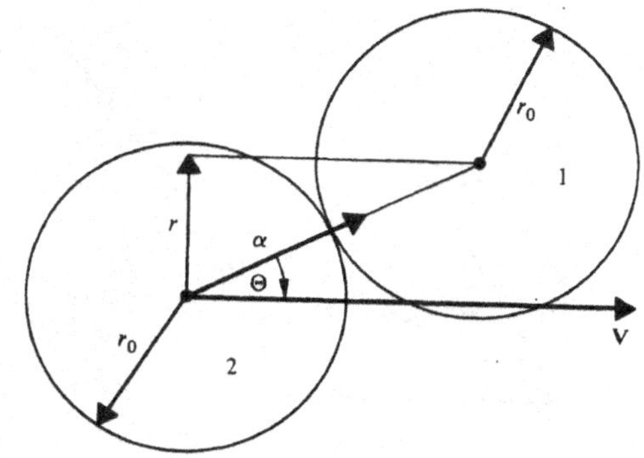

Figure 3–5 Configuration at impact for hard sphere molecules.

impact parameter r, as shown, is the distance between lines drawn through the centers of particles 1 and 2 parallel to **V** (which in this case coincides with the asymptotic trajectory). The obvious geometrical relationship $r = 2r_0 \sin \theta$ allows us to immediately determine $B(\theta,V)$:

$$B(\theta,V) = Vr\left|\frac{\partial r}{\partial \theta}\right| = 4r_0^2\, V \sin \theta \cos \theta \tag{3–34}$$

3–4 GENERAL INTERMOLECULAR POTENTIALS

For intermolecular potentials other than the hard sphere model the relationship between Θ and r is not immediately transparent. In this section we will consider general results, which we then specialize in the next section for the case of power law potentials. Our problem is most easily treated by casting it into the form of another problem, that of the motion of a single particle in a central force field—one of the classical problems of particle dynamics. Let us then first formulate this latter problem, and then show how the results obtained can be applied to the former problem.

Consider the motion of a particle of mass M moving in a central field of force. The force on the particle at any time is directed on a line defined by the particle and some fixed point in space, the force center. The trajectory of the particle will be confined to the plane defined by the radius vector **R**, drawn from the force center to the particle, and the instantaneous velocity vector of the particle, since all the forces exerted on the particle are directed in this plane. The two planar coordinates which define the position of the particle are **R** and the angle ψ (see Figure 3–6). The kinetic energy of the particle can be expressed in terms of

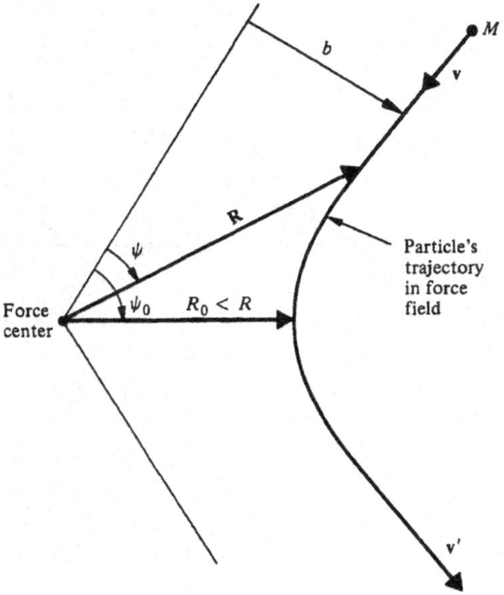

Figure 3–6

these coordinates as follows:

$$
\begin{aligned}
\mathrm{KE} &= \frac{M}{2}(\dot{x}^2 + \dot{y}^2) \\
&= \frac{M}{2}(\dot{R}^2 + R^2\dot{\psi}^2) \\
&\equiv \frac{M}{2}\left(p_R{}^2 + \frac{p_\psi{}^2}{M^2R^2}\right)
\end{aligned}
\tag{3-35}
$$

The quantity $\mathbf{p}_\psi$ is the particle's angular momentum, which is a constant of the motion and can therefore be replaced by its value prior to the particles entering the force field, $M\mathbf{v}b$. If $\phi(R)$ is the potential from which we assume the central force is derivable, the total energy of the particle along its trajectory E is

$$
\begin{aligned}
E &= \text{kinetic energy} + \text{potential energy} \\
&= \frac{M}{2}\left[\dot{R}^2 + \left\{\frac{vb}{R}\right\}^2\right] + \phi(R) = \frac{M}{2}v^2
\end{aligned}
\tag{3-36}
$$

thus

$$
\frac{\dot{R}^2}{v^2} = 1 - \frac{b^2}{R^2} - \frac{\phi(R)}{\frac{M}{2}v^2}
$$

Since $\dot{\psi}^2 = (bv/R^2)^2$, we can then determine the equation for the particle orbit as follows:

$$\left(\frac{dR}{d\psi}\right)^2 = \left(\frac{\dot{R}}{\dot{\psi}}\right)^2 = \frac{R^4}{b^2}\left\{1 - \frac{b^2}{R^2} - \frac{\phi(R)}{\frac{M}{2}v^2}\right\}$$

so that

$$d\psi = \pm \frac{b}{R^2}\frac{dR}{\left[1 - \frac{b^2}{R^2} - \frac{\phi(R)}{\frac{1}{2}Mv^2}\right]} \tag{3-37}$$

Along the approach trajectory $(dR/d\psi)$ is clearly negative so that we have, finally,

$$\psi_0 = \int_0^{\psi_0} d\psi = -\int_\infty^R \frac{dR\, b}{R^2\left[1 - \left(\frac{b}{R}\right)^2 - \frac{\phi(R)}{\frac{M}{2}v^2}\right]^{1/2}}$$

$$= \int_0^{\nu_0} \frac{d\nu}{\left[1 - \nu^2 - \frac{2\phi(b\nu^{-1})}{Mv^2}\right]^{1/2}} \tag{3-38}$$

where the upper limit, ν_0, is the positive root of the equation

$$1 - \nu_0^2 - \frac{2\phi(b\nu_0^{-1})}{Mv^2} = 0 \tag{3-39}$$

Note that the angle ψ_0 bears the same relationship to the angle through which $\mathbf{v}$ is turned as does Θ to the angle through which $\mathbf{V}$ is turned in the binary collision problem.

Let us now see how the above results can be used to describe the binary collision parameters which appear in the Boltzmann collision integral. For a two-particle system undergoing a classical binary collision, we define the quantities $\mathbf{R}_c$, $\mathbf{R}_r$ as follows:

$$\mathbf{R}_c = \frac{(m_1\mathbf{q}_1 + m_2\mathbf{q}_2)}{2(m_1 + m_2)}$$

$$= \frac{(\mathbf{q}_1 + \mathbf{q}_2)}{2}, \quad (m_1 = m_2 \equiv m)$$

$$\mathbf{R}_r = \mathbf{r}_{12} = \mathbf{q}_1 - \mathbf{q}_2 \tag{3-40}$$

The quantity $\mathbf{R}_c$ defines the center of mass of the two-particle system, and the quantity $\mathbf{R}_r$ is the interparticle separation. The relative velocity and acceleration are therefore $\dot{\mathbf{R}}_r$ and $\ddot{\mathbf{R}}_r$, and the equation of motion for each particle follows from Newton's law:

$$m\ddot{\mathbf{q}}_1 = -\frac{\partial \phi_{12}}{\partial \mathbf{q}_1}$$
$$m\ddot{\mathbf{q}}_2 = -\frac{\partial \phi_{12}}{\partial \mathbf{q}_2} = \frac{\partial \phi_{12}}{\partial \mathbf{q}_1} \tag{3-41}$$

As a consequence of the above equations of motion it follows that $\ddot{\mathbf{R}}_c = 0$, that is, the center of mass of the system moves with constant velocity. A further consequence of these equations is that

$$\ddot{\mathbf{R}}_r = -\left(\frac{1}{m} + \frac{1}{m}\right)\frac{\partial \phi_{12}}{\partial \mathbf{R}_r}$$
$$\frac{m}{2}\ddot{\mathbf{R}}_r = -\frac{\partial \phi_{12}}{\partial \mathbf{R}_r} \tag{3-42}$$

We see that the relative motion takes place in a plane which contains the center of mass and the vectors $\dot{\mathbf{R}}_c$, $\ddot{\mathbf{R}}_c$. In fact, the above equation is just the equation of motion for a single particle of mass $m/2$ moving in a central field characterized by the potential ϕ_{12}. Thus we can carry over our earlier results in toto by replacing M with $(\frac{1}{2})m$, and $\mathbf{v}$ by $\mathbf{V}$. Further, as we mentioned earlier, ω_0 is replaced by Θ, and b, the distance of closest approach, is replaced by r. The complete transcription is summarized in the table below.

TABLE 3–1

PARTICLE IN A CENTRAL FIELD		CLASSICAL BINARY COLLISION
$\phi(R)$		$\phi_{12}(R_r)$
M		$\frac{m}{2}$
$\mathbf{v}$	$\Rightarrow$	$\mathbf{V}$
ψ_0		θ
b		r

We will therefore have the following relationship for the classical binary collision parameters:

$$\Theta = \int_0^{\nu_0} \frac{d\nu}{\left\{1 - \nu^2 - \dfrac{4\phi_{12}(r\nu^{-1})}{mV^2}\right\}^{1/2}}$$
$$1 - \nu_0^2 - \frac{4\phi_{12}(r\nu_0^{-1})}{mV^2} = 0 \tag{3-43}$$

3-5 POWER LAW POTENTIALS

The results of the previous section can be made more explicit when the intermolecular potential is specified. A general class of potentials which has been extensively used in kinetic theory is the so-called power law potential,

$$\phi_{12}(R_r) \propto \frac{1}{R_r^{s-1}}, \qquad s > 3 \tag{3-44}$$

The hard sphere potential is the limiting case of the power law potential for $s \to \infty$, $K^{1/s-1} \to 2r_0$ where we denote the proportionality constant indicated above as K. The expression which we obtained in the last section for Θ, Equation (3-43), can now be written as

$$\Theta = \int_0^{\nu_0} \frac{d\nu}{\left\{1 - \nu^2 - \dfrac{4K\nu^{s-1}}{mV^2r^{s-1}}\right\}^{1/2}} \tag{3-45}$$

or

$$\Theta(\beta) = \int_0^{\nu_0} \frac{d\nu}{\{1 - (\nu/\beta)^{s-1} - \nu^2\}^{1/2}} \tag{3-46}$$

where $\beta = (mV^2/4K)^{1/s-1}r$. We can thus explicitly determine $B(\Theta,V)$, which for this case will be

$$\begin{aligned} B(\Theta,V) &= Vr\left|\frac{dr}{d\theta}\right| \\ &= V\left(\frac{4K}{m}\right)^{2/s-1} \frac{\beta}{V^{4/s-1}} \frac{d\beta}{d\Theta} \\ &= \left(\frac{4K}{m}\right)^{2/s-1} V^{(s-5/s-1)}\beta\frac{d\beta}{d\Theta} \end{aligned} \tag{3-47}$$

where $\beta(\Theta)$ can be found by inversion from $\theta(\beta)$. The term $B\,d\theta\,d\epsilon$ which appears in the collision integral can now be replaced by

$$\left(\frac{2K}{m}\right)^{2/s-1} V^{(s-5/s-1)}\beta\,d\beta\,d\epsilon$$

We see that for the case $s = 5$ we have $B(\Theta,V) = B(\Theta)$ which, as we shall see in later chapters, leads to considerable simplifications in the theory of the Boltzmann equation. This result was first obtained by Maxwell, and molecules which are supposed to interact in this manner are called Maxwell molecules.

In concluding this section we will briefly consider the asymptotic behavior of $\Theta(\beta)$ and explicitly show why the potential must be generally cut off. First consider the case of small β. We have

$$\Theta_{\beta\to 0} \approx \int_0^{\beta} \frac{d\nu}{\left\{1 - \left(\frac{\nu}{\beta}\right)^{s-1}\right\}^{1/2}}$$

$$= \beta \int_0^1 \frac{du}{\{1 - u^{s-1}\}^{1/2}} \tag{3-48}$$

and we see that in this limit Θ is linear in β and well behaved. For the other limiting case, β large, we have

$$\Theta_{\beta\to\infty} \approx \int_0^{\nu_0} \frac{d\nu}{\{\nu_0^2 - \nu^2 + (\nu_0/\beta)^{s-1} - (\nu/\beta)^{s-1}\}^{1/2}}$$

$$\approx \int_0^1 \frac{du}{(1 - u^2)^{1/2}} \left(1 - \frac{1}{2\beta^{s-1}} \frac{1 - u^{s-1}}{1 - u^2}\right)$$

$$= \frac{\pi}{2} - \frac{A(s)}{\beta^{s-1}} \tag{3-49}$$

We then have, for $\beta \to \infty$,

$$\beta \propto \left(\frac{\pi}{2} - \theta\right)^{-1/s-1}$$

or

$$\beta^2 \propto \left(\frac{\pi}{2} - \theta\right)^{-2/s-1}$$

so that

$$\beta \frac{d\beta}{d\theta} \propto \left(\frac{\pi}{2} - \theta\right)^{-(s+1/s-1)} \tag{3-50}$$

which shows that $B(\Theta,V)$ contains a nonintegrable singularity at $\theta = \frac{1}{2}\pi$. This can be avoided either by cutting off β, so that the potential ϕ_{12} is zero for $\beta > \beta_{max} \ll \infty$, or by the less physical, but mathematically more tractable, method of directly cutting off Θ near $\frac{1}{2}\pi$, that is, eliminating grazing collisions from the collision term (a grazing collision is one in which there is little change in the direction of the relative velocity). Such difficulties do not, of course, occur in the hard sphere model, in which the potential is explicitly of a finite range.

3-6 THE COLLISION CROSS SECTION

We will conclude this chapter by establishing some relationships which we will find useful in later chapters. Let us begin by introducing the differential collision cross section, $\sigma(\chi,V)$, which we define in terms of the deflection angle $\chi = \pi - 2\theta$ as follows:

$$\sigma(\chi,V) \sin \chi \, d\chi \, d\epsilon \equiv \sigma(\chi,V) \, d\Omega = r \, dr \, d\epsilon \qquad (3\text{-}51)$$

The quantity $d\Omega = \sin \chi \, d\chi \, d\epsilon$ is the differential solid angle at angle χ with the line $r = 0$ (Figure 3-7). Although $\sigma(\chi,V)$ may be given a physical

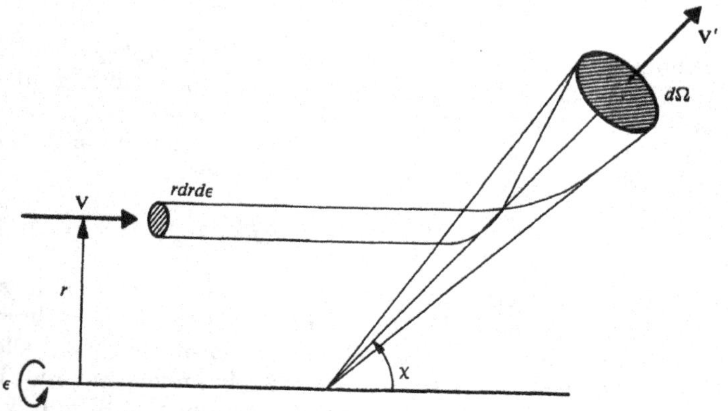

Figure 3-7

interpretation in terms of particle beam scattering, for our purposes it is preferable to consider it a defined quantity. We can determine σ directly by expressing r (or β) as a function of χ instead of Θ, so that we have

$$\sigma(\chi,V) = \frac{r}{\sin \chi} \frac{dr}{d\chi} \qquad (3\text{-}52)$$

As we will see later, integrals of the form

$$\begin{aligned} \phi_{12}^{(l)} &\equiv \int_0^\pi d\chi \sin \chi \, \sigma(\chi,V) V(1 - \cos^l \chi) \\ &= \int_0^\infty dr \, rV(1 - \cos^l \chi) \end{aligned} \qquad (3\text{-}53)$$

play an important role in the theory of Boltzmann's equation. The notation is standard, and $\phi_{12}^{(l)}$ is not to be confused with $\phi_{12}(R_r)$, the intermolecular potential. Let us consider the evaluation of the above integral for a power law potential. Using the results obtained in the preceding section, we see that

$$\phi_{12}^{(l)} = \left(\frac{2K}{m}\right)^{2/s-1} V^{(s-5/s-1)} \int_0^\infty d\beta\, \beta(1 - \cos^l \chi\,(\beta))$$
$$= \left(\frac{2K}{m}\right)^{2/s-1} V^{(s-5/s-1)} A_l(s) \tag{3-54}$$

where

$$A_l(s) = \int_0^\infty d\beta\, \beta(1 - \cos^l \chi)$$

is a pure number which can be calculated by substituting Equation (3–46) for $\chi(\beta)$. This quantity has been computed, and the results are given in Chapter 8 for the various significant values of l and s.

The Boltzmann collision integral can be written in terms of the differential collision cross section, in which case we have

$$J(f) = \frac{1}{m}\int d\mathbf{v}_2\, d\Omega\, \sigma(\chi,V)V[f_1'f_2' - f_1f_2] \tag{3-55}$$

where the primed velocities are now to be considered as functions of χ and ϵ. We have introduced the above form of $J(f)$ in order to show that it is possible to write this term in still another form, which will allow us to prove, with minimum difficulty, an extremely important result in the next chapter. If we postulate the existence of a quantity $W(\mathbf{v}_1,\mathbf{v}_2/\mathbf{v}_1',\mathbf{v}_2')$, which we define to be the probability that a binary collision changes the velocities of particles 1 and 2 from $\mathbf{v}_1$, $\mathbf{v}_2$ to $\mathbf{v}_1'$, $\mathbf{v}_2'$, then it is perhaps intuitively clear that $J(f)$ can be written as

$$J(f) = \frac{1}{m}\int d\mathbf{v}_2\, d\mathbf{v}_1'\, d\mathbf{v}_2'\, \{W(\mathbf{v}_1',\mathbf{v}_2'/\mathbf{v}_1,\mathbf{v}_2)f_1'f_2' - W(\mathbf{v}_1,\mathbf{v}_2/\mathbf{v}_1',\mathbf{v}_2')f_1f_2\} \tag{3-56}$$

In order to rigorously demonstrate the above form for $J(f)$ we must first determine what properties the W's must have. Since W cannot depend on the numbering of the particles, it must satisfy the following symmetry property:

$$W(\mathbf{v}_1,\mathbf{v}_2/\mathbf{v}_1',\mathbf{v}_2') = W(\mathbf{v}_2,\mathbf{v}_1/\mathbf{v}_2',\mathbf{v}_1') \tag{3-57}$$

Further, the probability of a direct collision must be the same as that of the inverse collision, since these events are in one-to-one correspondence. Therefore, W must also satisfy the following inversion property:

$$W(\mathbf{v}_1,\mathbf{v}_2/\mathbf{v}_1',\mathbf{v}_2') = W(\mathbf{v}_1',\mathbf{v}_2'/\mathbf{v}_1,\mathbf{v}_2) \tag{3-58}$$

This last relationship allows us to write the collision term in the following form:

$$J(f) = \frac{1}{m}\int d\mathbf{v}_2\, d\mathbf{v}_1'\, d\mathbf{v}_2'\, W(\mathbf{v}_1,\mathbf{v}_2/\mathbf{v}_1',\mathbf{v}_2')[f_1'f_2' - f_1f_2] \tag{3-59}$$

In the above representation the collision term contains a nine-fold integration, as opposed to the five-fold integration which appears in the usual representations, for example, Equation (3–55). In order to bring $J(f)$ as written above into a standard form we must consider a proper reduction of the dimensionality of the integration which appears in the above integral form. This can be done by the more detailed consideration of the structure of the W functions which follows.

There are three independent quantities which W depends on. We may choose these, as above, to be $\mathbf{v}_2$, $\mathbf{v}_1'$, $\mathbf{v}_2'$ ($\mathbf{v}_1$ is a parameter) or, as we now elect, to be $\mathbf{v}_1 - \mathbf{v}_2$, $\mathbf{v}_1' - \mathbf{v}_2'$, and $\mathbf{v}_1 + \mathbf{v}_2 - \mathbf{v}_1' - \mathbf{v}_2'$. As we have seen, the center of mass of a system of two colliding particles has a constant velocity during the collision. Changing this quantity for a particular collision corresponds to a transformation of the coordinate system in which that collision is described, and does not therefore change the probability of that collision. Thus, we can then write

$$\begin{aligned} W(\mathbf{v}_1,\mathbf{v}_2/\mathbf{v}_1',\mathbf{v}_2') &= W(\mathbf{v}_2 - \mathbf{v}_1, \mathbf{v}_2' - \mathbf{v}_1', \mathbf{v}_1 + \mathbf{v}_2 - \mathbf{v}_1' - \mathbf{v}_2') \\ &= W(\mathbf{v}_2 - \mathbf{v}_1, \mathbf{v}_2' - \mathbf{v}_1')\delta(\mathbf{v}_1 + \mathbf{v}_2 - \mathbf{v}_1' - \mathbf{v}_2') \end{aligned} \qquad \textbf{(3–60)}$$

where the delta function is necessary to insure that all allowable collisions satisfy conservation of momentum. To proceed further, we again change the independent variables, using this time $\mathbf{V}'$ and $\mathbf{V}_c' \equiv \frac{1}{2}(\mathbf{v}_1' + \mathbf{v}_2')$ instead of $\mathbf{v}_1'$, $\mathbf{v}_2'$. Since $d\mathbf{v}_1'\,d\mathbf{v}_2' = d\mathbf{V}_c'\,d\mathbf{V}'$, inserting Equation (3–60) into $J(f)$ and carrying out the $\mathbf{V}_c'$ integration gives us the six-dimensional integral

$$J(f) = \frac{1}{m}\int d\mathbf{v}_2\,d\mathbf{V}'\,W(\mathbf{V},\mathbf{V}')[f_1'f_2' - f_1f_2] \qquad \textbf{(3–61)}$$

where, as shown below, the primed velocities are now explicit in the independent variables,

$$\begin{aligned} \mathbf{v}_1' &= \mathbf{V}_c + \tfrac{1}{2}\mathbf{V}' \\ \mathbf{v}_2' &= \mathbf{V}_c - \tfrac{1}{2}\mathbf{V}' \end{aligned} \qquad \textbf{(3–62)}$$

The angle between $\mathbf{V}$ and $\mathbf{V}'$ has already been defined as χ. Since the collision probability is independent of the absolute spatial orientation of the two colliding particles, we can therefore express $W(\mathbf{V},\mathbf{V}')$ in terms of χ and $V = V'$. A delta function must again be introduced into the resulting expression for W to insure that the proper conservation law, in this case Equation (3–5), is not violated by an allowable collision. We then have

$$W(\mathbf{V},\mathbf{V}') = \sigma'(\chi,V)\delta\left(\frac{V^2 - V'^2}{2}\right) \qquad \textbf{(3–63)}$$

Writing

$$d\mathbf{V}' = \sin\chi\,d\chi\,d\epsilon\,V'^2\,dV' = d\Omega\,V'^2\,dV'$$

(see Figure 3–7), so that

$$\int_0^\infty dV'\, V'^2\delta\left(\frac{V^2 - V'^2}{2}\right) = \int_0^\infty d\left(\frac{V'^2}{2}\right) V'\delta\left(\frac{V'^2 - V^2}{2}\right) = V$$

we get, finally,

$$J(f) = \frac{1}{m}\int d\mathbf{V}\, d\Omega\, \sigma'(\chi,V)V[f_1'f_2' - f_1f_2] \qquad \textbf{(3–64)}$$

Comparing the above result with Equation (3–55), we see that we can identify $\sigma'(\chi,V) = \sigma(\chi,V)$ with

$$\sigma(\chi,V) = \sigma\left(\arccos\frac{\mathbf{V}\cdot\mathbf{V}'}{VV'}, V\right) \qquad \textbf{(3–65)}$$

We have thus proved the result we set out to, namely that $J(f)$ can be represented by the form given by Equation (3–56), where

$$W(\mathbf{v}_1,\mathbf{v}_2/\mathbf{v}_1',\mathbf{v}_2') = \sigma\left(\arccos\frac{\mathbf{V}\cdot\mathbf{V}'}{\mathbf{V}\mathbf{V}'}, V\right)\delta(\mathbf{v}_1 + \mathbf{v}_2 - \mathbf{v}_1' - \mathbf{v}_2')\,\delta\left(\frac{V^2 - V'^2}{2}\right) \qquad \textbf{(3–66)}$$

Although we will not use the above form for $J(f)$ in calculations, it will prove useful, as we shall see in the next chapter, in proving some special properties of the linearized collision integral.

Reference

Most of the material here is standard and can be found in any of the standard references mentioned earlier. The use of the W functions in the formulation of $J(f)$ is due to Waldmann (1958). For another method of writing $J(f)$ see:

1. L. Finkelstein, *Phys. of Fluid*, vol. 8, p. 431, 1965.

Problems

3–1. Prove that $d\mathbf{v}_1\, d\mathbf{v}_2 = d\mathbf{v}_1'\, d\mathbf{v}_2'$ by calculating the Jacobian of the transformation.

3–2. Verify Equations (3–4) and (3–5).

3–3. Suppose that instead of structureless molecules, we consider molecules with internal degrees of freedom and associated internal energies. The conservation of energy equation for a binary collision then takes the form

$$v_1^2 + v_2^2 = v_1'^2 + v_2'^2 + \frac{2\,\Delta E}{m}$$

where ΔE is the change in internal energy for the collision. Derive equations similar to Equations (3–4) and (3–5) for this case, and interpret them geometrically.

3–4. Consider a one-dimensional system of particles which interact only through binary collisions. Explicitly, solve Equations (3–2) and (3–3) for the post-collisional velocities in terms of the pre-collisional velocities, and explain, for the special case of a space-uniform system, why we would no longer need a Boltzmann equation to describe the system.

3–5. Calculate $\nu(\mathbf{v}_1)$ for particles which interact through a cutoff Maxwell molecule potential.

3–6. Calculate $\nu(\mathbf{v}_1)$ for hard sphere molecules when the system is in a state for which

$$f = \frac{\rho}{(2\pi RT)^{3/2}}\, e^{-\mathbf{v}^2/2RT}$$

3–7. Rigorously prove that the hard sphere potential can be considered as a power law potential.

3–8. Calculate the differential collision cross section $\sigma(\chi,V)$ for hard sphere molecules.

3–9. For general power law potentials $B(\Theta,V)$ will diverge for $\Theta \to \pi/2$. Explain why this divergence does not generally occur for a hard sphere potential, and then show that in fact this divergence is implicit in the value of $B(\Theta,V)$ calculated by us for hard spheres [Equation (3–34)].

3–10. The Boltzmann collision integral contains a five-fold integration. Prove that for a two-dimensional gas the Boltzmann collision integral will contain a three-fold integration.

3–11. Determine $W(\mathbf{v}_1',\mathbf{v}_2'/\mathbf{v}_1,\mathbf{v}_2)$ for a two-dimensional system, and use this result to obtain the three-fold collision integral for this case.

CHAPTER 4

properties of the collision term

4–1 SYMMETRY PROPERTIES OF $J(f)$

In this section we will make use of the results of Section 3–1 to derive some important symmetry properties of $J(f)$. These properties, which will be used repeatedly in the following chapters, concern the various representations of the expression

$$\int d\mathbf{v}_1\, J(f)\psi(\mathbf{v}_1) = \frac{1}{m}\int d\mathbf{v}_1\, d\mathbf{v}_2\, d\epsilon\, d\theta\, B(\theta,V)[f_1'f_2' - f_1f_2]\psi(\mathbf{v}_1) \quad (4\text{–}1)$$

where $\psi(\mathbf{v}_1)$ is *any* arbitrary function of $\mathbf{v}_1$.

It is clear that a simple interchange of the dummy variables $\mathbf{v}_1$ and $\mathbf{v}_2$ in the integrand on the right-hand side of Equation (4–1), which produces [see Equations (3–1) and (3–8)] an interchange of the dependent variables $\mathbf{v}_1'$ and $\mathbf{v}_2'$ will give us the following form for the integral:

$$\int d\mathbf{v}_1\, J(f)\psi(\mathbf{v}_1) = \frac{1}{m}\int d\mathbf{v}_1\, d\mathbf{v}_2\, d\epsilon\, d\theta\, B(\theta,V)[f_1'f_2' - f_1f_2]\psi(\mathbf{v}_2) \quad (4\text{–}2)$$

Further, replacing the dummy variables $\mathbf{v}_1$, $\mathbf{v}_2$ by $\mathbf{v}_1'$, $\mathbf{v}_2'$, which, from Equations (3–7), (3–8), (3–11), and (3–12), we see leads to a transformation of the dependent variables $\mathbf{v}_1'$, $\mathbf{v}_2'$ to $\mathbf{v}_1$, $\mathbf{v}_2$, gives us the following alternate form for this integral:

$$\begin{aligned}\int d\mathbf{v}_1\, J(f)\psi(\mathbf{v}_1) &= -\frac{1}{m}\int d\mathbf{v}_1'\, d\mathbf{v}_2'\, d\epsilon\, d\theta\, B(\theta,V')[f_1'f_2' - f_1f_2]\psi(\mathbf{v}_1') \\ &= -\frac{1}{m}\int d\mathbf{v}_1\, d\mathbf{v}_2\, d\epsilon\, d\theta\, B(\theta,V)[f_1'f_2' - f_1f_2]\psi(\mathbf{v}_1') \quad (4\text{–}3)\end{aligned}$$

where we have used Equations (3–5) and (3–18) to obtain the second equation. The identical replacements as in the right-hand side of Equation

(4–2) gives us one more representation,

$$\int d\mathbf{v}_1\, J(f)\psi(\mathbf{v}_1) = -\frac{1}{m}\int d\mathbf{v}_1\, d\mathbf{v}_2\, d\epsilon\, d\theta\, B(\theta,V)[f_1'f_2' - f_1f_2]\psi(\mathbf{v}_2') \tag{4–4}$$

Combining Equations (4–1) through (4–4), we have, finally,

$$\begin{aligned}\int d\mathbf{v}_1\, J(f)\psi(\mathbf{v}_1) &= \frac{1}{4m}\int d\mathbf{v}_1\, d\mathbf{v}_2\, d\epsilon\, d\theta\, B(\theta,V)[f_1'f_2' - f_1f_2] \\ &\quad \cdot [\psi(\mathbf{v}_1) + \psi(\mathbf{v}_2) - \psi(\mathbf{v}_1') + \psi(\mathbf{v}_2')] \\ &= \frac{1}{4}\int d\mathbf{v}_1\, J(f)[\psi(\mathbf{v}_1) + \psi(\mathbf{v}_2) - \psi(\mathbf{v}_1') - \psi(\mathbf{v}_2')]\end{aligned} \tag{4–5}$$

The above result expresses the basic symmetry property of $J(f)$.

4–2 THE SUMMATIONAL INVARIANTS

The last result which we showed in the preceding section can be used to prove the statement embodied in Equation (2–24), which we originally stated without proof. We have, from Equation (4–5),

$$\int d\mathbf{v}_1\, J(f)\begin{bmatrix}1\\ \mathbf{v}_1\\ v^2\end{bmatrix} = \frac{1}{4}\int d\mathbf{v}_1\, J(f)\begin{bmatrix}1 + 1 - 1 - 1\\ \mathbf{v}_1 + \mathbf{v}_2 - \mathbf{v}_1' - \mathbf{v}_2'\\ v_1^2 + v_2^2 - v_1'^2 - v_2'^2\end{bmatrix} = 0 \tag{4–6}$$

where equality with zero is a consequence of the conservation of number of particles, momentum, and energy, respectively, in a binary collision. Equation (4–5) shows that $\int d\mathbf{v}_1\, J(f)\psi(\mathbf{v}_1)$ can be written in terms of the net change in the quantity $\psi(\mathbf{v}_1) + \psi(\mathbf{v}_2)$ due to a binary collision. Any molecular property $\psi(\mathbf{v}_1)$ for which there is no net change due to a collision, that is, so that

$$\begin{aligned}\int d\mathbf{v}_1\, J(f)\psi(\mathbf{v}_1) &= \frac{1}{2}\int d\mathbf{v}_1\, J(f)[\psi(\mathbf{v}_1) + \psi(\mathbf{v}_2)] \\ &= \frac{1}{4}\int d\mathbf{v}_1\, J(f)[\psi(\mathbf{v}_1) + \psi(\mathbf{v}_2) - \psi(\mathbf{v}_1') - \psi(\mathbf{v}_2')] = 0\end{aligned} \tag{4–7}$$

is called a summational invariant. It is intuitively clear that the number of particles, $(\psi(\mathbf{v}_1) = 1)$, the linear momentum $(\psi(\mathbf{v}_1) = m\mathbf{v}_1)$, and the total energy $(\psi(\mathbf{v}_1) = v_1^2/2m)$ are included in this category of functions. What is not obvious is the statement that these quantities are the only summational invariants. Let us now turn to the proof of this important result. The proof which follows is that given by Desloge;[1] for alternate proofs, see Kennard or Grad (1949).

Any summational invariant $\hat{\psi}(\mathbf{v}_1)$ must satisfy the generalized conservation equation

$$\hat{\psi}(\mathbf{v}_1) + \hat{\psi}(\mathbf{v}_2) = \hat{\psi}(\mathbf{v}_1') + \hat{\psi}(\mathbf{v}_2') \tag{4-8}$$

We will take the above equation to be the strict definition of a summational invariant. Note that this definition is not equivalent to one which might be based on Equation (4–7), although any function $\hat{\psi}(\mathbf{v}_1)$ that satisfies the above definition also satisfies that equation; the converse, however, is not necessarily true. Introducing now the new variables $\mathbf{W}$, $\mathbf{W}'$, where

$$\mathbf{W} = \tfrac{1}{2}(\mathbf{v}_1 + \mathbf{v}_2) = \mathbf{W}' = \tfrac{1}{2}(\mathbf{v}_1' + \mathbf{v}_2') \tag{4-9}$$

and the unit vectors $\mathbf{n}$ in the $\mathbf{V}$ direction and $\mathbf{n}'$ in the $\mathbf{V}'$ direction, we have the following relationships:

$$\begin{aligned} \mathbf{v}_2 &= \mathbf{W} + \tfrac{1}{2}\mathbf{V} = \mathbf{W} + \tfrac{1}{2}V\mathbf{n} \\ \mathbf{v}_1 &= \mathbf{W} - \tfrac{1}{2}\mathbf{V} = \mathbf{W} - \tfrac{1}{2}V\mathbf{n} \\ \mathbf{v}_2' &= \mathbf{W}' + \tfrac{1}{2}\mathbf{V}' = \mathbf{W} + \tfrac{1}{2}V\mathbf{n}' \\ \mathbf{v}_1' &= \mathbf{W}' - \tfrac{1}{2}\mathbf{V}' = \mathbf{W} - \tfrac{1}{2}V\mathbf{n}' \end{aligned} \tag{4-10}$$

We can use the above relationships to rewrite Equation (4–8) in the following form:

$$\hat{\psi}(\mathbf{W} - \tfrac{1}{2}V\mathbf{n}) + \hat{\psi}(\mathbf{W} + \tfrac{1}{2}V\mathbf{n}) = \hat{\psi}(\mathbf{W} - \tfrac{1}{2}V\mathbf{n}') + \hat{\psi}(\mathbf{W} + \tfrac{1}{2}V\mathbf{n}') \tag{4-11}$$

Let us now differentiate the above equation twice with respect to $\mathbf{V}$, which gives

$$\begin{aligned} \frac{\partial^2\hat{\psi}}{\partial V_\alpha \partial V_\beta}(\mathbf{W} - \tfrac{1}{2}V\mathbf{n}) + \frac{\partial^2\hat{\psi}}{\partial V_\alpha\, \partial V_\beta}(\mathbf{W} + \tfrac{1}{2}V\mathbf{n}) &= \frac{\partial^2\hat{\psi}}{\partial V_\alpha\, \partial V_\beta}(\mathbf{W} - \tfrac{1}{2}V\mathbf{n}') \\ &\quad + \frac{\partial^2\hat{\psi}}{\partial V_\alpha\, \partial V_\beta}(\mathbf{W} + \tfrac{1}{2}V\mathbf{n}') \end{aligned} \tag{4-12}$$

or, if $\mathbf{v}_1$ and $\mathbf{v}_2$ are considered as the independent variables,

$$\begin{aligned} n_\alpha n_\beta &\left\{\frac{\partial^2\hat{\psi}}{\partial v_{1_\alpha}\, \partial v_{1_\beta}}(\mathbf{W} - \tfrac{1}{2}V\mathbf{n}) + \frac{\partial^2\hat{\psi}}{\partial v_{2_\alpha}\, \partial v_{2_\beta}}(\mathbf{W} + \tfrac{1}{2}V\mathbf{n})\right\} \\ &\equiv n_\alpha n_\beta\{\hat{\psi}_{\alpha\beta}(\mathbf{W} - \tfrac{1}{2}V\mathbf{n}) + \hat{\psi}_{\alpha\beta}(\mathbf{W} + \tfrac{1}{2}V\mathbf{n})\} \\ &= n_\alpha' n_\beta'\{\hat{\psi}_{\alpha\beta}(\mathbf{W} - \tfrac{1}{2}V\mathbf{n}') + \hat{\psi}_{\alpha\beta}(\mathbf{W} + \tfrac{1}{2}V\mathbf{n}')\} \end{aligned} \tag{4-13}$$

[1] Professor Desloge has informed me that he posed the question to a mathematician colleague who brought the problem to the attention of his associates, one of whom (whose name, unfortunately, has been forgotten) subsequently devised the proof given here.

Here, aside from the indicated notational simplification, we have again made use of the standard summation convention; repeated indices, which will always be Greek subscripts in this book, are to be summed over.

The above equation is completely general and therefore is valid for all $\mathbf{V}$ and $\mathbf{W}$, including $V = 0$, $W = v_1$, in which case we have

$$n_\alpha n_\beta \hat{\psi}_{\alpha\beta}(\mathbf{v}_1) = n_\alpha' n_\beta' \hat{\psi}_{\alpha\beta}(\mathbf{v}_1) \tag{4–14}$$

The generality of Equation (4–13) requires that this equation must also be satisfied for all $\mathbf{n}$, $\mathbf{n}'$. The choices $\mathbf{n} = \hat{\mathbf{i}}$, $\mathbf{n}' = \hat{\mathbf{j}}$, and $\mathbf{n} = \hat{\mathbf{i}}$, $\mathbf{n}' = (\frac{1}{2})^{1/2}(\hat{\mathbf{i}} + \hat{\mathbf{j}})$, when substituted into Equation (4–14), give the following relationships:

$$\hat{\psi}_{11} = \hat{\psi}_{22} \tag{4–15}$$

$$\hat{\psi}_{11} = \frac{\hat{\psi}_{11}}{2} + \hat{\psi}_{12} + \frac{\hat{\psi}_{22}}{2} \tag{4–16}$$

from which we see that $\hat{\psi}_{12} = 0$. This procedure can be continued and the following general result is arrived at:

$$\hat{\psi}_{\alpha\beta}(\mathbf{v}_1) = \hat{\psi}(\mathbf{v}_1)\delta_{\alpha\beta} \tag{4–17}$$

Since $\hat{\psi}_{\alpha\beta} = 0$, $\alpha \neq \beta$, it follows that $\hat{\psi}_\alpha$ is a function of $v_{1\alpha}$ only. Therefore, from Equation (4–17) we have

$$\hat{\psi}_{11}(\mathbf{v}_1) = \hat{\psi}_{22}(\mathbf{v}_1) = \hat{\psi}_{33}(\mathbf{v}_1) = A \tag{4–18}$$

and integrating this expression gives

$$\hat{\psi}_\alpha(\mathbf{v}_1) = A v_{1\alpha} + B_\alpha \tag{4–19}$$

where A and B_α are constants. Let us now write $\hat{\psi}(\mathbf{v}_1) = \hat{\psi}(v_{11},v_{12},v_{13})$ and consider the following identity:

$$\begin{aligned}
\int_0^{v_{11}} dv_{11}' \, \hat{\psi}_1(v_{11}',0,0) &+ \int_0^{v_{12}} dv_{12}' \, \hat{\psi}_2(v_{11},v_{12}',0) + \int_0^{v_{13}} dv_{13}' \, \hat{\psi}(v_{11},v_{12},v_{13}') \\
&= \{\hat{\psi}(v_{11},0,0) - \hat{\psi}(0)\} + \{\hat{\psi}(v_{11},v_{12},0) - \hat{\psi}(v_{11},0,0)\} \\
&\qquad + \{\hat{\psi}(\mathbf{v}_1) - \hat{\psi}(v_{11},v_{12},0)\} \\
&= \hat{\psi}(\mathbf{v}_1) - \hat{\psi}(0)
\end{aligned} \tag{4–20}$$

The above integrals can be evaluated explicitly by substituting into them the result given by Equation (4–19) for the $\hat{\psi}_\alpha$; doing this we get, finally,[2]

$$\hat{\psi}(\mathbf{v}_1) = A\tfrac{1}{2}m\mathbf{v}_1^2 + \mathbf{B} \cdot m\mathbf{v}_1 + C \tag{4–21}$$

with C a constant. Thus we see that the most general form for any summational invariant is a linear combination of the summational invariants particle number, momentum, and total energy, which is the result we set out to prove.

[2] When $\mathbf{v}_1^2$ appears it denotes $\mathbf{v}_1 \cdot \mathbf{v}_1$ (scalar product).

4-3 THE LINEARIZED COLLISION TERM, $L(g)$

In addition to the Boltzmann collision term $J(f)$, some of whose properties we have already considered, we will also find it necessary to consider the linearized collision integral, which we now define as

$$L(g) \equiv \frac{1}{m}\int d\mathbf{v}_2\, d\epsilon\, d\theta\, B(\theta,V)(g_2' + g_1' - g_1 - g_2)f_M(\mathbf{v}_2) \tag{4-22}$$

where $f_M(\mathbf{v}_2)$ is the Maxwellian distribution function

$$f_M(\mathbf{v}_1) = \frac{\rho}{(2\pi RT)^{3/2}}\, e^{-(1/2RT)(\mathbf{v}_1-\mathbf{u})^2} \tag{4-23}$$

In the next chapter we will see that f_M is the most general value of f that satisfies $J(f) = 0$. Further, f_M with ρ, $\mathbf{u}$, T constant[3] characterizes the equilibrium state of the system. The quantity $f_M(\mathbf{v}_1)L(g)$ will appear in the Chapman–Enskog theory of the solution to the Boltzmann equation, while $L(g)$ alone appears in the theory of the linearized Boltzmann equation, which we obtain by setting $f = f_M(1 + g)$, substituting into the Boltzmann equation, and retaining only those terms linear in the pertubation $g \ll 1$, so that we have

$$\frac{\partial g}{\partial t} + \mathbf{v}_1 \cdot \frac{\partial g}{\partial \mathbf{q}} = L(g) \tag{4-24}$$

Let us consider some of the important properties of $L(g)$. The same operations which lead to Equation (4–5) also will allow us to obtain an analogous result for $f_M(\mathbf{v}_1)L(g)$,

$$\begin{aligned}\int d\mathbf{v}_1\, f_M(\mathbf{v}_1)L(g)\psi(\mathbf{v}_1) &= \frac{1}{4}\int d\mathbf{v}_1\, f_M(\mathbf{v}_1)L(g)[\psi(\mathbf{v}_1) + \psi(\mathbf{v}_2) - \psi(\mathbf{v}_1') - \psi(\mathbf{v}_2')]\\ &= \frac{1}{4m}\int d\mathbf{v}_1\, d\mathbf{v}_2\, d\theta\, d\epsilon\, B(\theta,V)[g_1' + g_2' - g_1 - g_2][\psi_1 + \psi_2 - \psi_1' - \psi_2']\\ &\qquad \cdot f_M(\mathbf{v}_1)f_M(\mathbf{v}_2)\end{aligned} \tag{4-25}$$

from which it follows immediately that $f_M(\mathbf{v}_1)L(g)$ is symmetric (self-adjoint), that is,

$$\int d\mathbf{v}_1\, f_M(\mathbf{v}_1)L(g)\psi(\mathbf{v}_1) = \int d\mathbf{v}_1\, f_M(\mathbf{v}_1)L(\psi)g(\mathbf{v}_1) \tag{4-26}$$

and nonpositive,

[3] In what follows it will not be necessary to distinguish between the cases ρ, $\mathbf{u}$, T constant and ρ, $\mathbf{u}$, T functions of $\mathbf{q}$ and t, and we use f_M for both. In the next chapter we will use f_M only for the case where ρ, $\mathbf{u}$, T are constant.

$$\int d\mathbf{v}_1 f_M(\mathbf{v}_1)L(g)g(\mathbf{v}_1) = -\frac{1}{4m}\int d\mathbf{v}_1\, d\mathbf{v}_2\, d\epsilon\, d\theta\, B(\theta,V)$$
$$[g_1' + g_2' - g_1 - g_2]^2 f_M(\mathbf{v}_1) f_M(\mathbf{v}_2)$$
$$\leq 0 \tag{4-27}$$

It is of interest to inquire into the properties of the eigenvalues λ_i, of the operator L. These are defined in the usual way, so that we will have

$$L(\psi_i) = \lambda_i\psi_i(\mathbf{v}_1) \tag{4-28}$$

From Equation (4–2) it follows that

$$\lambda_i = \frac{\int d\mathbf{v}_1 f_M(\mathbf{v}_1)L(\psi_i)\psi_i(\mathbf{v}_1)}{\int d\mathbf{v}_1 f_M(\mathbf{v}_1)\psi_i^2(\mathbf{v}_1)}$$
$$= -\frac{\int d\mathbf{v}_1\, d\mathbf{v}_2\, d\epsilon\, d\theta\, B(\theta,V)[\psi_{i_1}' + \psi_{i_2}' - \psi_{i_1} - \psi_{i_2}]^2 f_M(\mathbf{v}_1) f_M(\mathbf{v}_2)}{4m\int d\mathbf{v}_1 f_M(\mathbf{v}_1)\psi_{i_1}^2}$$
$$\leq 0 \tag{4-29}$$

Thus the eigenvalues λ_i are nonpositive. The eigenvalue zero is a common (degenerate) eigenvalue of the eigenfunctions 1, $\mathbf{v}_1$, $\mathbf{v}_1^2$ corresponding to the summational invariants, and all other eigenvalues are thus negative. It may also be easily verified that the ψ_i are orthogonal in the sense that

$$\int d\mathbf{v}_1 f_M(\mathbf{v}_1)\psi_i(\mathbf{v}_1)\psi_j(\mathbf{v}_1) = \delta_{ij}\int d\mathbf{v}_1 f_M(\mathbf{v}_1)\psi_i^2(\mathbf{v}_1) \tag{4-30}$$

The eigenvalues λ_i can be interpreted as reciprocal relaxation times characteristic of Maxwell molecules. To see this we consider the initial value problem associated with a small spatially uniform disturbance in a system initially in equilibrium. We will have $g = g(\mathbf{v}_1,t)$, and if we assume that the eigenfunctions ψ_i are discrete and form a complete set, which as we will see is a plausible assumption for Maxwell molecules, we can then expand g in the ψ_i. Doing this, so that we have

$$g = \sum_i \alpha_i(t)\psi_i(\mathbf{v}_1)$$

substituting into the linearized Boltzmann equation, and making use of the orthogonality properties of the ψ_i, leads to the following result:

$$g(\mathbf{v}_1,t) = \sum_i \alpha_i e^{\lambda_i t}\psi_i(\mathbf{v}_1)$$

with α_i a constant. The $\lambda_i < 0$ are clearly the set of relaxation times which characterize the problem, while the zero eigenvalues give rise to nondecaying changes in the first five moments, density, macroscopic flow velocity, and temperature.

4-4 AN ALTERNATE REPRESENTATION OF $L(g)$

We now consider the proof of a result which will play an important role in our later considerations. This result, simply stated, is that for suitably cutoff potentials we can write $f_M(\mathbf{v}_1)L(g)$ in the following form:

$$f_M(\mathbf{v}_1)L(g) = -\nu_L(\mathbf{v}_1)f_M(\mathbf{v}_1)g(\mathbf{v}_1) + \int d\mathbf{v}_2\, K(\mathbf{v}_1,\mathbf{v}_2)g(\mathbf{v}_2) \quad \textbf{(4-31)}$$

where the kernel $K(\mathbf{v}_1,\mathbf{v}_2)$ is symmetric, that is, $K(\mathbf{v}_1,\mathbf{v}_2) = K(\mathbf{v}_2,\mathbf{v}_1)$.

The reader familiar with the theory of integral equations will recognize that the above results imply that the integral equation

$$f_M(\mathbf{v}_1)L(g) = 0 \quad \textbf{(4-32)}$$

is thus representable as a homogeneous Fredholm integral equation of the second kind, an identity we will later exploit in making use of the properties of such equations. For the time being, however, we wish only to show that Equation (4–31) is, in fact, valid. This result is not immediately obvious; it was originally proved by Hilbert for the special case of hard spheres, and in complete generality by Enskog and later, more concisely, by Grad. The proof that follows is due to Waldmann, and makes use of the alternate form of the collision term, $J(f)$, given by Equation (3–59). The linearized version of this form gives rise to the following expression for $L(g)$:

$$f_M(\mathbf{v}_1)L(g) = \frac{1}{m}\int d\mathbf{v}_2\, d\mathbf{v}_1'\, d\mathbf{v}_2'\, W(\mathbf{v}_1,\mathbf{v}_2/\mathbf{v}_1',\mathbf{v}_2')[f_M(\mathbf{v}_1)f_M(\mathbf{v}_2)f_M(\mathbf{v}_1')f_M(\mathbf{v}_2')]^{1/2} \cdot [g_1' + g_2' - g_1 - g_2] \quad \textbf{(4-33)}$$

where we have made use of the conservation laws for momentum and energy in writing $f_M(\mathbf{v}_1)f_M(\mathbf{v}_2) = [f_M(\mathbf{v}_1)f_M(\mathbf{v}_2)f_M(\mathbf{v}_1')(f_M(\mathbf{v}_2')]^{1/2}$. In the above equation, the term that contains g_1 can be written as

$$\frac{1}{m}\int d\mathbf{v}_2\, d\mathbf{v}_1'\, d\mathbf{v}_2'\, f_M(\mathbf{v}_1)f_M(\mathbf{v}_2)W(\mathbf{v}_1,\mathbf{v}_2/\mathbf{v}_1',\mathbf{v}_2')g_1 = \nu_L(\mathbf{v}_1)f_M(\mathbf{v}_1)g(\mathbf{v}_1) \quad \textbf{(4-34)}$$

where, from Equation (3–66),

$$\nu_L(\mathbf{v}_1) = \frac{1}{m}\int d\mathbf{v}_2\, d\epsilon\, d\theta\, B(\theta,V)f_M(\mathbf{v}_2) \quad \textbf{(4-35)}$$

Note that for cutoff Maxwell molecules $B(\theta,V)$, and hence ν_L, is a constant independent of $\mathbf{v}_1$. With the appropriate changes of variable the remaining terms in Equation (4–33) can be written as

$$\int d\mathbf{v}_2 \left[\int d\mathbf{v}_1'\, d\mathbf{v}_2' \,\{W(\mathbf{v}_1,\mathbf{v}_1'/\mathbf{v}_2,\mathbf{v}_2') + W(\mathbf{v}_1,\mathbf{v}_1'/\mathbf{v}_2',\mathbf{v}_2) - W(\mathbf{v}_1,\mathbf{v}_2/\mathbf{v}_1',\mathbf{v}_2')\} \cdot \{f_M(\mathbf{v}_1)f_M(\mathbf{v}_2)f_M(\mathbf{v}_1')f_M(\mathbf{v}_2')\}^{1/2}\right] g(\mathbf{v}_2) = \int d\mathbf{v}_2\, K(\mathbf{v}_1,\mathbf{v}_2)g(\mathbf{v}_2) \qquad (4\text{–}36)$$

The symmetry of the kernel K follows immediately from the symmetry and inversion properties of the W functions. A further consequence of these properties is that

$$\nu_L(\mathbf{v}_1) = \int d\mathbf{v}_2\, K(\mathbf{v}_1,\mathbf{v}_2) \qquad (4\text{–}37)$$

Collecting the above results, we see that we have the form for $f_M(\mathbf{v}_1)L(g)$ given by Equation (4–31).

It can be shown, using sophisticated mathematical techniques beyond the scope of this book, that for appropriately cutoff potentials the continuous eigenvalue spectrum of L is given by the values taken on by $\nu_L(\mathbf{v}_1)$. Thus, for molecules other than Maxwell molecules, there is a continuous spectrum, and, accordingly, the eigenfunctions do not form a discrete set. For Maxwell molecules, however, as we noted above, ν_L is a constant, and for this singular case the eigenfunctions form a discrete set, so that, as indicated in the previous section, an expansion of g in this set is plausible.

4–5 THE EIGENTHEORY OF $L(g)$

As we have pointed out in the preceding section, the properties of the eigenspectrum of the linearized collision operator depend quite sensitively on the interparticle potential. In general, the eigenfunctions of L will be of the form

$$\psi_i(\mathbf{v}) = \psi_{rlm}(\mathbf{v}) = \psi_{rl}(v)Y_{lm}(\theta_v,\phi_v) \qquad (4\text{–}38)$$

where $Y_{lm}(\theta_v,\phi_v)$ are spherical harmonics defined over the velocity space. This follows from the fact that L is isotropic in the velocity space (that is, L commutes with rotation operators on the velocity space). For general interparticle potentials the eigenvalue $\lambda = 0$ corresponds to those eigenfunctions which are summational invariants. We will always refer to these as ψ_i $(i = 1 - 5)$, where

$$\begin{aligned} \psi_1 &= 1 \\ \psi_{2,3,4} &= v_{x,y,z}(RT)^{-1/2} \\ \psi_5 &= \left(\frac{3}{2}\right)^{1/2}\left(1 - \frac{v^2}{3RT}\right) \end{aligned} \qquad (4\text{–}39)$$

according to the normalization

$$\int d\mathbf{v}\, \frac{e^{-\mathbf{v}^2/2RT}}{(2\pi RT)^{3/2}}\, \psi_i \psi_j = \delta_{ij} \tag{4–40}$$

For interparticle potentials other than Maxwell molecules, the precise details of the remaining part of the eigenspectrum have yet to be determined. In the case of Maxwell molecules, however, an exact theory is available, since in this case $B(\theta,V) = B(\theta)$, which considerably simplifies the problem. Recall that the transformations [Equations (3–11) and (3–12)] which take the primed velocities into $\mathbf{v}_1$, $\mathbf{v}_2$ are linear, so that if $\psi(\mathbf{v})$ is a polynomial of given degree in $\mathbf{v}$, then $L(\psi)$ will also be a polynomial of the same degree. Of course, this is only true when B is independent of V. We can therefore conclude that for Maxwell molecules the eigenfunctions of L will be polynomials. The determination of the exact form of these eigenfunctions is quite complicated. The proof was first given in the Appendix of an unpublished report by Wang Chang and Uhlenbeck, and later, independently, by Waldmann in his *Handbuch* article. Their results are:

$$\psi_{rl}(v) = \left(\frac{\pi^{1/2} r!}{2(l + \frac{1}{2} + r)!}\right)^{1/2} S_{l+1/2}{}^{(r)}\left(\frac{v^2}{RT}\right)\left(\frac{v}{\sqrt{RT}}\right)^l \tag{4–41}$$

and

$$\lambda_{rl} = 2\pi \int_0^{\pi/2} d\theta\, B(\theta)\{1 - \sin^{l+2r}\theta[P_l(\sin\theta) - \cos^{l+2r}\theta]P_l(\cos\theta)\}$$

where $S_{l+1/2}{}^{(r)}$ are the Sonnine polynomials which are discussed in Chapter 6, and the P_l are Legendre polynomials.

References

The proof in Section 4–2 can be found in:

1. E. A. Desloge, *Statistical Physics*. New York: Holt, Rinehart and Winston, 1966.

For alternate proofs see:

2. H. Grad, *Commun. Pure and Appl. Math.*, vol. 2, p. 331, 1949.
3. E. H. Kennard, *Kinetic Theory of Gases*. New York: McGraw-Hill, 1938.

The theory of the linearized collision operator was developed by:

4. T. Carleman (1957). (See General References.)

5. C. Cercignani, *Phys. Fluids*, vol. 10, p. 2097, 1967.

6. H. Grad, in *Rarefied Gas Dynamics*, J. Laurmann, Ed. New York: Academic Press, 1963.

7. D. Hilbert, *Math. Ann.*, vol. 72, p. 562, 1912.

The results for the eigenfunctions of L were first given (for Maxwell molecules) by:

8. L. Waldmann (1958).

9. C. S. Wang Chang and G. E. Uhlenbeck, "On the propagation of sound in monatomic gases," Engineering Research Institute, University of Michigan, Ann Arbor, Project M999, October 1952 (unpublished).

Problems

4–1. Prove that the quantity $\int d\mathbf{v}\, J(f) \ln f$ can never be positive. For what specific values of f will this quantity be zero?

4–2. Show that $\int d\mathbf{v}\, f_M(\mathbf{v}) g(\mathbf{q},\mathbf{v},t) L(g)$ can never be positive. For what specific values of g will this quantity be zero?

4–3. We define the bilinear form of the collision operator as $J(\ ,\ )$, where

$$J(f,g) \equiv \frac{1}{2m} \int d\mathbf{v}_2\, d\theta\, d\epsilon\, B(\theta,V)\{f_1'g_2' + f_2'g_1' - f_1g_2 - f_2g_1\}$$

Express $L(g)$ and $J(f)$ in terms of $J(\ ,\ )$.

4–4. Prove that the operator L has only five eigenfunctions corresponding to the eigenvalue zero.

4–5. What physical phenomena would be associated with positive eigenvalues of L, and what does the nonexistence of such eigenvalues therefore imply?

4–6. Prove the orthogonality relationship of Equation (4–30).

4–7. If f is weakly perturbed away from

$$f_M = \frac{\rho_0}{(2\pi RT_0)^{3/2}}\, e^{-\mathbf{v}^2/(2RT_0)}$$

so that $f = f_M(1 + g)$, then accordingly we will have $\rho = \rho_0(1 + \tilde{\rho})$, $\mathbf{u} = \tilde{\mathbf{u}}$, $T = T_0(1 + \tilde{T})$. Show that if we expand g in the eigenfunctions of L according to

$$g(\mathbf{q},\mathbf{v},t) = \sum_i c_i(\mathbf{q},t)\psi_i(\mathbf{v})$$

then c_1, $c_{2,3,4}$, and c_5 are given by $\bar{\rho}$, $u_{x,y,z}$ and $-(\frac{3}{2})^{1/2}T$, respectively.

4–8. Explain why $\nu_L(\mathbf{v}_1)$ as defined in Equation (4–35) is identical to the quantity $\nu(\mathbf{v}_1)$ defined in Equation (3–29) for Maxwell molecules. Why won't this identity generally be true for other power law potentials?

4–9. Explicitly calculate $\nu_L(\mathbf{v}_1)$ for hard sphere molecules, and show that this quantity is a monatone increasing function of $\mathbf{v}_1$. The quantity $\nu_L(\mathbf{v}_1)$ can be interpreted as a velocity dependent collision frequency. With ν_L viewed in this light, what simple physical interpretation can be given to its monotonicity?

4–10. Determine the constants α_i (which appear in the equation for $g(\mathbf{v},t)$ given at the end of Section 3) in terms of $g(\mathbf{v},0)$. Simplify the equation for $g(\mathbf{v},t)$ in the case where $g(\mathbf{v},0)$ is a nondegenerate eigenfunction of L.

CHAPTER 5

the *H* theorem and irreversibility

5-1 SPATIALLY HOMOGENEOUS SYSTEMS

On the basis of our preliminary considerations we can already anticipate that solving Boltzmann's equation will be an arduous task, since it is both nonlinear and has an integro-differential form (and is decidedly nonstandard). Boltzmann himself was unable to obtain explicit solutions of this equation. However, even without solving it, he was able to extract a great deal of significant information from it. We have already seen how it is possible, without solving Boltzmann's equation, to obtain the usual hydrodynamical equations from it. Another important result we can obtain from Boltzmann's equation, without knowing its solution, is the H theorem, a result first proved by Boltzmann, which we now proceed to consider.

Let us introduce the following two functionals of the Boltzmann equation solution, f:

$$\begin{aligned} H[f] &= H(\mathbf{q},t) \equiv \int d\mathbf{v}_1\, f_1 \ln f_1 \\ \bar{H}[f] &= \bar{H}(t) \equiv \int d\mathbf{q}\, d\mathbf{v}_1\, f_1 \ln f_1 \end{aligned} \tag{5-1}$$

In the present section we will consider systems which are spatially homogeneous, that is,

$$f(\mathbf{q},\mathbf{v},t) = \rho f(\mathbf{v},t)$$

where ρ is now constant throughout the system.

When the system is in an equilibrium state, f is stationary, that is, $(\partial f/\partial t) = 0$ and therefore $\partial H/\partial t$ is also zero. Our physical intuition, based on observation, tells us that an isolated system will evolve from an arbitrary initial state to a state of equilibrium. Boltzmann's H theorem formalizes this notion, and also makes explicit the manner in which this evolution proceeds. Further, Boltzmann's result allows us to deduce the form of f in the equilibrium state; in our later considerations we will make repeated use of this aspect of the H theorem.

Before proceeding we must consider an important property of the H function, namely that it cannot assume all negative values. To see this we note that the total kinetic energy of the system must be finite, so that

$$\int d\mathbf{v}_1 \, f_1 \mathbf{v}_1^2 < \infty \tag{5–2}$$

Now

$$H = \int d\mathbf{v}_1 \, f_1 \ln f_1$$

offers the possibility of a divergence since at large $\mathbf{v}_1$, although $f_1 \to 0$, $\ln f_1 \to -\infty$. However, the bound given by Equation (5–2) shows that if H does diverge then $\ln f_1$ must approach infinity faster than v_1^2, or, equivalently, f_1 approaches zero faster than $e^{-v_1^2}$. If this is the case, however, then H must converge since $\lim_{x\to\infty} e^{-x^2}x^n = 0$ for all values of n. Thus H cannot decrease without bound.

The H theorem states that H can only decrease with increasing time. To demonstrate this result we consider

$$\frac{\partial H}{\partial t} = \frac{\partial}{\partial t}\int d\mathbf{v}_1 \, f_1 \ln f_1$$

$$= \int d\mathbf{v}_1 \, J(f) \ln f_1 \tag{5–3}$$

where we have made use of the fact that $\int d\mathbf{v}_1 \, f_1$ is time invariant. Making use of the symmetry properties of the collision term [see Equation (4–5)] we can rewrite Equation (5–3) in the following form:

$$\frac{\partial H}{\partial t} = \frac{1}{4}\int d\mathbf{v}_1 \, J(f)\{\ln f_1 + \ln f_2 - \ln f_1' - \ln f_2'\}$$

$$= -\frac{1}{4}\int d\mathbf{v}_1 \, d\mathbf{v}_2 \, d\theta \, d\epsilon \, B(\theta,V)\{f_1 f_2 - f_1' f_2'\} \ln \frac{f_1 f_2}{f_1' f_2'} \tag{5–4}$$

The cross section $B(\theta,V)$ is positive for all values of its arguments, so we can conclude that

$$\frac{\partial H}{\partial t} \le 0 \tag{5–5}$$

since in the above integral representation the logarithm always takes on the same sign as the term in the brackets. The equality, which holds in equilibrium, is obtained only for states for which

$$f_1 f_2 = f_1' f_2' \tag{5–6}$$

and these are thus the only equilibrium states of the system.

The H theorem tells us that the system approaches equilibrium in a strictly monatone manner, and further, specifies the equilibrium state. The equilibrium state is characterized by Equation (5–6), or equivalently,

$$\ln f_1 + \ln f_2 = \ln f_1' + \ln f_2' \tag{5-7}$$

The above description of the equilibrium state indicates that the $\ln f$ in this state is a summational invariant [see Equation (4–8)], so that

$$\ln f_M(\mathbf{v}_1) = A + \mathbf{B} \cdot \mathbf{v}_1 + C\mathbf{v}_1^2 \tag{5-8}$$

The five constants which appear in the above expression are not arbitrary, since we have an equal number of constraints relating the macroscopic to the first five moments of f. Leaving the determination of these quantities as an exercise, we quote the resulting unique value of f obtained for the equilibrium state:

$$f_M(\mathbf{v}_1) = \frac{\rho}{(2\pi RT)^{3/2}} e^{-(\mathbf{v}_1-\mathbf{u})^2/2RT} \tag{5-9}$$

The subscript M is used to denote the equilibrium distribution function since that quantity is usually referred to as the Maxwellian distribution function (sometimes Maxwell–Boltzmann distribution function). In the, present case, where ρ, $\mathbf{u}$, T are space independent quantities, f_M is referred to as an absolute Maxwellian to distinguish it from the case, considered in the next section, where these quantities may depend on space and time. The equilibrium form for f obtained above from the H theorem is identical to that obtained independently using the methods of equilibrium statistical mechanics. This provides an additional consistency test on the methods of the nonequilibrium theory which we are considering herein.

Having determined the explicit form of f in equilibrium we can calculate many of the gas properties in this state. Of particular interest is the pressure tensor

$$\mathsf{P} = \int d\mathbf{v}_0\, \mathbf{v}_0\mathbf{v}_0 \frac{\rho}{(2\pi RT)^{3/2}} e^{-\mathbf{v}_0^2/2RT} \tag{5-10}$$

which is seen to be diagonal and symmetric. Evaluating the above integral, we verify the perfect gas law $p = \rho RT$. The heat flux vector, $\mathbf{Q}$, can also be calculated; this quantity is trivially shown to be zero in equilibrium.

5–2 THE H THEOREM FOR A NONUNIFORM SYSTEM

Let us now consider a spatially nonuniform system, and inquire as to whether we can still expect that there will be a monatone approach to a unique equilibrium state. Multiplying Boltzmann's equation by

($\log f_1 + 1$) and integrating over $\mathbf{q}$ and $\mathbf{v}_1$ leads to the following equation:

$$\frac{\partial \bar{H}}{\partial t} + \oint_{\substack{\text{system}\\ \text{boundary}}} d\mathbf{S} \cdot \int d\mathbf{v}_1\, \mathbf{v}_1 f_1 \ln f_1 = \int d\mathbf{q}\, d\mathbf{v}_1\, J(f) \ln f_1 \qquad (5\text{–}11)$$

We now assume that there is no flux of H at the system boundaries. This condition is realized if the system boundaries are such that we have a specular reflection of particles at the walls, so that, for example, at any portion of the boundary normal to the x axis of our reference coordinate system

$$f(\mathbf{q},v_x,v_y,v_z,t) = f(\mathbf{q}, -v_x, v_y,v_z,t) \qquad \mathbf{q} \text{ on } d\mathbf{S} \perp x \qquad (5\text{–}12)$$

In this case the surface integral which appears in Equation (5–11) vanishes and we find, similar to our earlier result,

$$\frac{\partial \bar{H}}{\partial t} = \int d\mathbf{q}\, d\mathbf{v}_1\, J(f) \ln f_1 \leq 0 \qquad (5\text{–}13)$$

The unique solution for f which leads to the equality statement in the above expression is once more of the Maxwellian form, but now ρ, $\mathbf{u}$, and T can be functions of space and time. In this case we say that f is locally Maxwellian. Since a locally Maxwellian distribution leads to a vanishing collision term, the left-hand side of Boltzmann's equation must also vanish when f assumes this particular form (we denote the local Maxwellian by the subscript LM), that is,

$$\frac{\partial f_{\text{LM}}}{\partial t} + \mathbf{v}_1 \cdot \frac{\partial f_{\text{LM}}}{\partial \mathbf{q}} = J(f_{\text{LM}}) = 0 \qquad (5\text{–}14)$$

We must now inquire what additional conditions are imposed by the left-hand side of the above equation. Of the several solutions to this equation, the most general will be shown below to be the absolute Maxwellian. Other solutions are also possible, but these are only permissible with special boundary conditions, and are not often encountered in practice. Accordingly, we will not treat these special equilibrium distribution functions in any detail.

The solution to Equation (5–14) must satisfy $J(f_{\text{LM}}) = 0$, so that we must require

$$\ln f_{\text{LM}} = A + \mathbf{B} \cdot \mathbf{v}_1 + C \frac{\mathbf{v}_1^2}{2m} \qquad (5\text{–}15)$$

where A, $\mathbf{B}$, and C are now possibly functions of space and time. Substituting the above expression into the left-hand side of Equation (5–14), and equating the coefficients of like powers of $\mathbf{v}_1$ to zero (which appears as the common coefficient of all powers of $\mathbf{v}_1$ on the right-hand side of the equation) we have[1]

$$\begin{aligned} \frac{\partial A}{\partial t} &= 0 \\ \frac{\partial \mathbf{B}}{\partial t} + \frac{\partial}{\partial \mathbf{q}} A &= 0 \\ \frac{\partial C}{\partial t}\delta_{ij} + \frac{1}{2}\left(\frac{\partial B_i}{\partial q_j} + \frac{\partial B_j}{\partial q_i}\right) &= 0 \\ \frac{\partial C}{\partial \mathbf{q}} &= 0 \end{aligned} \tag{5–16}$$

To satisfy the above equations the coefficients must be of the following form:

$$\begin{aligned} A &= A_1 + \mathbf{A}_2 \cdot \mathbf{q} + A_{3_{\alpha\beta}} q_\alpha q_\beta \\ B_i &= B_{1_i} + B_{2_i} t + B_{3_{i\alpha}} q_\alpha + B_{4_{i\alpha}} q_\alpha t \\ C &= C_1 + C_2 t + C_3 t^2 \end{aligned} \tag{5–17}$$

Substituting these relationships back into Equation (5–16) we then obtain

$$\begin{aligned} A &= A_1 + \mathbf{A}_2 \cdot \mathbf{q} + C_3 q^2 \\ \mathbf{B} &= \mathbf{B}_1 - \mathbf{A}_2 t - (2C_3 t + C_2)\mathbf{q} + \boldsymbol{\Omega} \wedge \mathbf{q} \\ C &= C_1 + C_2 t + C_3 t^2 \end{aligned} \tag{5–18}$$

where the vector $\boldsymbol{\Omega}$ is defined below.

$$\boldsymbol{\Omega} = B_{3_{kj}} \hat{\imath} + B_{3_{ik}} \hat{\jmath} + B_{3_{ji}} \hat{k} \tag{5–19}$$

The choice $\mathbf{A}_2 = C_2 = C_3 = \boldsymbol{\Omega} = 0$ implies an absolute Maxwellian distribution function, with ρ, $\mathbf{u}$, T constant. If some of the above constants are not zero, then other forms for the equilibrium distribution function are possible, but as mentioned above, these cases are of little general interest and will not be considered here.

We have seen that for a spatially uniform system the approach to equilibrium is strictly monatone, the equilibrium state being characterized by f assuming the absolute Maxwellian form. For a nonequilibrium system we expect that there will be a monatone approach to equilibrium in the velocity space, that is, to a state in which f is locally Maxwellian with ρ, $\mathbf{u}$, T varying in space and time. This state is approached, but not reached, due to interference from the streaming part (left-hand side) of the Boltzmann equation. (This conjecture has been proved for the

[1] Remember that the summation convention only applies to repeated Greek indices.

linearized Boltzmann equation by Grad.) The approach to equilibrium in the coordinate space should not necessarily be monatone, and in general will take much longer than the equilibration in the velocity space.

5–3 IRREVERSIBILITY AND BOLTZMANN'S EQUATION

The quantity $H[f_M]$ can be explicitly evaluated, leading to the following relationship:

$$H[f_M] = \int d\mathbf{v}_1 \, f_M \log f_M = -\rho \frac{S}{R} + \text{constant} \qquad (5\text{–}20)$$

Here S is the thermodynamic entropy density for a perfect gas. Thus, in the BGL we see that the H function (actually $-H$) represents a sort of generalized thermodynamic entropy for nonequilibrium states. The entropy increases (H decreases) according to the second law of thermodynamics until the equilibrium state consistent with the external constraints (in this case specification of ρ, $\mathbf{u}$, T) is reached, at which point the entropy (and H) remains constant. Thus the H theorem may be viewed as a microscopic version of the second law of thermodynamics *when the BGL is specified.*

The H theorem is one of the distinguishing features of the Boltzmann equation; the fact that this equation predicts a monatone (that is, irreversible) approach to equilibrium led to a great deal of criticism when this result was first published by Boltzmann. The basis for this criticism was that any description of a dynamical system based on Newton's laws must be, like those laws, reversible. We have already seen that Liouville's equation is reversible, and that

$$H_N = \int d\Gamma \, F_N \ln F_N$$

is constant for solutions of Liouville's equation. The critics argued that Boltzmann's equation must be wrong, since it furnishes an irreversible description, and consequently leads to violations of certain laws of mechanics. In particular, Zermelo cited the Poincare recurrence theorem which states that for a given closed mechanical system subject only to conservative forces, any state (that is, point in the Γ space) must recur to any desired accuracy (but not exactly) infinitely often. Thus H should return arbitrarily close to its initial value after a typical recurrence time, and cannot therefore uniformly decrease. A characteristic time for such recurrence for a dilute gas is $10^{10^{18}}$ seconds, by contrast with which we have the "age" of the universe estimated at 10^{17} seconds (thus Boltzmann's apocryphal answer to Zermelo—"you should wait so long"). Unfortunately, Boltzmann was not always at his best in answering his critics, and it was not until the Ehrenfests published their famous article in support of Boltzmann in 1911 that his ideas were put on a firm basis and widely accepted.

Let us briefly consider the essence of the Ehrenfests' argument in showing why Boltzmann's equation is not in conflict with the mechanical notions of reversibility. The important features of Boltzmann's equation overlooked by his critics are that, in addition to specifying *ab initio* a probabilistic rather than a mechanically deterministic description, so that we are always predicting expectations rather than certainties, we have further coarse grained (see below) and subsequently included an additional stochastic element into the description in the form of the *stosszahlansatz*. By coarse graining we mean that we have not considered the complete probabilistic description, which is given by F_N, but instead we have coarsened the description (coarse grained) by integrating out some of the variables, in this case $\mathbf{x}_2, \mathbf{x}_3, \cdots, \mathbf{x}_N$. Thus, in using F_1 we are using a very coarse description compared to that given by using F_N, and, in fact, we can show that due to coarse graining alone there will be a tendency for the resulting description to be irreversible.

Let us write F_N in the following form:

$$F_N(\Gamma,t) = \prod_{i=1}^{N} F_1(\mathbf{x}_i,t) + g(\Gamma,t) \tag{5–21}$$

where $g(\mathbf{x}_1,\mathbf{x}_2, \cdots ,\mathbf{x}_n,t)$ is the N-particle correlation function. This function must satisfy the condition

$$\int d\Gamma\, A(\mathbf{x}_i,t)g(\Gamma,t) = 0 \tag{5–22}$$

where $A(\mathbf{x}_i)$ is any function of the phase point $\mathbf{x}_i$ and t; this condition follows from Equation (5–21) and the normalization properties of F_N and F_1. We will want to make use of the mean value theorem in what follows. This theorem states that if a function $\mathfrak{F}(x + h)$ is expanded in a Taylor series in powers of h about $\mathfrak{F}(x)$ then

$$\mathfrak{F}(x + h) = \sum_{r=0}^{n} \frac{h^r}{r!} \mathfrak{F}^{(r)}(x) + R_n(\theta) \tag{5–23}$$

where the remainder term, $R_n(\theta)$, is given by

$$R_n(\theta) = \frac{h^{n+1}}{(n+1)!} \mathfrak{F}^{(n+1)}(x + \theta h) \qquad 0 \le \theta \le 1 \tag{5–24}$$

(the superscript indicates a derivative of the indicated order). Let us make use of this result to write

$$\begin{aligned}\ln F_N(\Gamma,t) &= \ln\left(\prod_{i=1}^{N} F_1(\mathbf{x}_i,t) + g(\Gamma,t)\right)\\ &= \ln \prod_{i=1}^{N} F_1(\mathbf{x}_i,t) + \frac{g(\Gamma,t)}{\prod_{i=1}^{N} F(\mathbf{x}_i,t) + \theta g(\Gamma,t)}\\ &= \sum_{i=1}^{N} \ln F_1(\mathbf{x}_i,t) + \frac{g(\Gamma,t)}{\prod_{i=1}^{N} F_1(\mathbf{x}_i,t) + \theta g(\Gamma,t)} \qquad (5\text{–}25)\end{aligned}$$

where we have specialized Equations (5–23) and (5–24) to the case $n = 0$. The above result now enables us to write

$$\begin{aligned}&\sum_{i=1}^{N} H_1(i,t) - H_N(1,2,\cdots N,t)\\ &\quad\equiv \sum_{i=1}^{N} \int d\mathbf{x}_i\, F_1(\mathbf{x}_i,t) \ln F_1(\mathbf{x}_i,t) - \int d\Gamma\, F_N(\Gamma,t) \ln F_N(\Gamma,t)\\ &\quad= -\int d\Gamma\; g(\Gamma,t)\left\{1 + \frac{(1-\theta)g(\Gamma,t)}{\prod_{i=1}^{N} F_1(\mathbf{x}_i,t) + \theta g(\Gamma,t)}\right.\\ &\quad= -\int \frac{d\Gamma\,(1-\theta)g^2(\Gamma,t)}{\prod_{i=1}^{N} F_1(\mathbf{x}_i,t) + \theta g(\Gamma,t)} \le 0 \qquad (5\text{–}26)\end{aligned}$$

The final inequality follows from the fact that the denominator which appears in the integrand is always positive, which can be seen from Equation (5–21), taking into account the positivity of F_N and F_1. Thus we can conclude that

$$H_N \ge \sum_{i=1}^{N} H_1(i,t) = NH_1(t) \qquad (5\text{–}27)$$

If we choose F_N so that initially it is a simple product of the F_1, that is, $g(\Gamma,0) = 0$, then we have $H_N = NH_1$ at $t = 0$. For times $t > 0$, H_N remains constant, and from Equation (5–27) we can conclude that H_1 can only decrease from its initial value. Thus the possibility of irreversibility has been introduced as a sole consequence of coarse graining. Irreversibility associated with a contraction in the level of description is also evident in the familiar macroscopic processes such as heat flow and diffusion. The macroscopic description is, of course, the coarsest possible.

A still stronger refutation of Boltzmann's critics is possible, for their arguments are predicted on the fact that the description contained in Boltzmann's equation is purely mechanical. However, the use of the *stosszahlansatz* has clearly introduced a nonmechanical element into this equation. This is a stochastic statement with no mechanical basis, and therefore Boltzmann's equation may be expected to lead to results contrary to what might be expected on the basis of purely mechanical arguments. Further, subtler points are brought out in the Ehrenfests' article and the more recent articles by Grad, and the reader who·wishes to be exposed to the fine points of this subject is encouraged to consult these sources.

References

For a discussion of the H theorem and the controversy surrounding it see:

1. L. Boltzmann (1964). See General References.
2. P. Ehrenfest and T. Ehrenfest, *The Conceptual Foundations of the Statistical Approach in Mechanics* (English translation by M. Moravosik). Ithaca, N.Y.: Cornell University Press, 1959.
3. H. Grad, *J. SIAM*, vol. 13, p. 259, 1965.
4. H. Grad, *Commun. Pure and Appl. Math.*, vol. 14, p. 323, 1961.
5. D. ter Haar, *Elements of Statistical Mechanics.* New York: Holt, Rinehart and Winston, 1954.

Problems

5–1. Consider $H[f]$ for a spatially uniform system, and show that the H function linearizes as

$$H = H[f_M] + \frac{1}{2}\int d\mathbf{v}\, f_M g^2 + O(g^3)$$

where the linearization is according to $f = f_M(1 + g)$.

5–2. Prove an H theorem for the linearized Boltzmann equation using the H function derived in Problem 5–1.

5–3. Show that the constraints on the moments of f, together with Equation (5–8) lead to the expression given in Equation (5–9) for f_M.

5–4. Calculate the extremum of the functional

$$H = \int d\mathbf{v}\, f \ln f$$

subject to the constraints

$$\int d\mathbf{v}\, f \begin{pmatrix} 1 \\ \mathbf{v} \\ v^2 \end{pmatrix} = \begin{pmatrix} \rho \\ \rho\mathbf{u} \\ 3RT + u^2 \end{pmatrix}$$

and thereby verify that the state for which H is a minimum is characterized by f_M.

5–5. When a system is close to equilibrium we would expect f and its derivatives with respect to time to be well behaved. Thus the approach to equilibrium should be characterized by the condition $\partial^2 H/\partial t^2 \geq 0$ in the linearized regime. Prove this for a space-uniform system.

5–6. Prove the inequality $H_N \geq H_n + H_{N-n}$, where

$$\begin{aligned} H_N &= \int d\Gamma_N\, F_N(\Gamma_N,t) \ln F_N(\Gamma_N,t) \\ H_{N-n} &= \int d\Gamma_{N-n}\, F_{N-n}(\Gamma_{N-n},t) \ln F_{N-n}(\Gamma_{N-n},t) \\ H_n &= \int d\Gamma_n\, F_n(\Gamma_n,t) \ln F_n(\Gamma_n,t) \end{aligned}$$

and

$$d\Gamma_N = \prod_{i=1}^{N} d\mathbf{x}_i, \qquad d\Gamma_n = \prod_{i}^{n} d\mathbf{x}_i, \qquad d\Gamma_{N-n} = \prod_{i=N-n}^{N} d\mathbf{x}_i$$

What is the condition for the equality to hold?

5–7. Quite often an irreversible physical process can be described in terms of a so-called master equation,

$$\dot{P}_i = \sum_j (A_{ji}P_j - A_{ij}P_i)$$

where the master function P_i is the probability of the state i of the process, A_{ij} is the transition probability per unit time from state i to j, and $P_i^0 A_{ij} = A_{ji}P_j^0$, where P_i^0 is the equilibrium value of P_i. Demonstrate an H theorem for the master equation using

$$H = \sum_i P_i \ln P_i/P_i^0$$

CHAPTER 6

normal solutions of the boltzmann equation

6–1 THE FREDHOLM THEOREMS

In this chapter we will make frequent use of several results from the theory of integral equations. We list these results below, so that we can conveniently refer to them when necessary.

As we have noted earlier, the equation

$$f_M(\mathbf{v}_1)L(g) = 0 \tag{6–1}$$

is a homogeneous Fredholm integral equation of the second kind. The more general equation,

$$f_M(\mathbf{v}_1)L(g) = h(\mathbf{q},\mathbf{v}_1,t) \tag{6–2}$$

in which h is a known function of its argument is the inhomogeneous Fredholm equation of the second kind associated with the homogeneous integral equation (6–1). If the associated homogeneous equation (6–1) has a solution

$$g(\mathbf{q},\mathbf{v}_1,t) = \sum_i a_i(\mathbf{q},t)\psi_i(\mathbf{v}_1) \tag{6–3}$$

then a solution to the inhomogeneous equation (6–2) will exist only if the following orthogonality conditions are satisfied:

$$\int d\mathbf{v}_1\, \psi_i(\mathbf{v}_1)h(\mathbf{q},\mathbf{v}_1,t) = 0 \qquad i = 0, 1, 2, \cdots \tag{6–4}$$

Further, if $\bar{g}(\mathbf{q},\mathbf{v}_1,t)$ is a particular solution of Equation (6–2), then it is unique providing the following conditions are satisfied:

$$\int d\mathbf{v}_1\, \psi_i(\mathbf{v}_1)\bar{g}(\mathbf{q},\mathbf{v}_1,t)f_M(\mathbf{v}_1) = 0 \qquad i = 0, 1, 2, \cdots \tag{6–5}$$

These are the Fredholm theorems, and we shall repeatedly refer to them in what follows.

6–2 HILBERT'S UNIQUENESS THEOREM

Let us consider the Boltzmann equation as an initial value problem. Then f is specified at some particular time, which we may take as $t = 0$, and Boltzmann's equation must be solved to find f for $t > 0$. In what follows we shall adopt the usual prerogative of the physical scientist and assume the existence of solutions to Boltzmann's equation; the actual proof of existence theorems is a matter of highly sophisticated mathematics which we will only briefly mention in a later chapter.

It is unrealistic to suppose that we would have so detailed a knowledge of any system so as to be able to specify f at any time for a nonequilibrium system. An equivalent description would also be given by specifying the complete (infinite) set of the moments of f, which is also clearly inaccessible to our knowledge. Thus the following result, due to the German mathematician Hilbert, takes on a very profound significance. If f can be expanded in powers of some small parameter, which we will denote by δ, then f is uniquely determined for $t > 0$ by the values at $t = 0$ of its first five moments $(\rho, \mathbf{u}, T)$ only. In other words, f is uniquely determined by the macroscopic thermo-fluid state. Such solutions of Boltzmann's equations are called normal solutions. A short proof of Hilbert's result is given below; this may be omitted at first reading.

The first step in Hilbert's proof is to note that the Boltzmann equation can be written with the factor $1/\delta$, $\delta \ll 1$, in front of the collision term in a particular physical regime. This is most easily seen by comparing the relative size of the terms which appear in the equation. If we let $\mathfrak{L}$ and $\mathfrak{T}$ denote, respectively, a typical macroscopic length and time, then the terms on the left-hand side of Boltzmann's equation are of order $\mathfrak{T}^{-1}$ and $u\mathfrak{L}^{-1}$, while the collision term is of order $(\rho/m)u\sigma^2$. Since the mean free path λ is proportional to $(\rho\sigma^2/m)^{-1}$ we can introduce still another characteristic time, $\bar{\mathfrak{T}} = \lambda u^{-1}$. Clearly $\bar{\mathfrak{T}}$ is a typical microscopic time; it is the time between successive collisions. Restricting ourselves now to the physical regime in which $\bar{\mathfrak{T}}\mathfrak{T}^{-1}$ is of the same order as $\lambda\mathfrak{L}^{-1}$, we then see that the collision term is of order $u\lambda^{-1}$, or of order $\mathfrak{L}\lambda^{-1}$ compared to the left-hand side of Boltzmann's equation. Introducing the parameter

$$\delta = \lambda\mathfrak{L}^{-1} \equiv K_n$$

we then see that in the regime in which the system behaves as a continuum (that is, where many collisions take place over a macroscopic length) δ is a small number. The ratio $\lambda\mathfrak{L}^{-1}$ is generally referred to as the Knudsen number, written K_n as shown above. It should be kept in mind that the solutions to Boltzmann's equation which we will consider are valid only for small Knudsen numbers.

Proceeding, we rewrite Boltzmann's equation in the form

$$\frac{\partial f}{\partial t} + \mathbf{v} \cdot \frac{\partial f}{\partial \mathbf{q}} \equiv Df = \frac{1}{\delta} J(f) \qquad (6\text{–}6)$$

and we look for solutions of the type

$$f = f(\mathbf{q},\mathbf{v},t;\delta) = \sum_{n=0}^{\infty} \delta^n f^{(n)}(\mathbf{q},\mathbf{v},t) \qquad (6\text{–}7)$$

Substituting the above expression back into Equation (6–6), and equating coefficients of like powers of δ yields the following set of equations for the $f^{(n)}$:

$$J(f^{(0)}) = 0 \qquad (6\text{–}8)$$

$$2J(f^{(0)},f^{(1)}) = Df^{(0)} \qquad (6\text{–}9)$$

$$\sum_{m=0}^{n} J(f^{(m)},f^{(n-m)}) = Df^{(n-1)} \qquad n > 1 \qquad (6\text{–}10)$$

Here we have introduced the so-called bilinear form of $J(f)$, $J(f^{(r)},f^{(s)})$, which is defined through the following relationship:

$$J(f_1^{(r)},f_1^{(s)}) \equiv \frac{1}{2m} \int d\mathbf{v}_2\, d\theta\, d\epsilon\, B(\theta,V) \cdot [f_1'^{(r)}f_2'^{(s)} + f_2'^{(r)}f_1'^{(s)} - f_1^{(r)}f_2^{(s)} - f_2^{(r)}f_1^{(s)}] \qquad (6\text{–}11)$$

Both $J(f)$ and $L(f)$ can be expressed in terms of the bilinear operator; we have $J(f) = J(f,f)$, and $f_M L(g) = 2J(f_M,f_M g)$.

The general unknown, $f^{(n)}$, is given by the equation

$$2J(f^{(0)},f^{(n)}) = Df^{(n-1)} - \sum_{m=1}^{n-1} J(f^{(m)},f^{(n-m)}) \qquad (6\text{–}12)$$

so that we have replaced our original problem of solving an integral-differential equation with the problem of solving a single homogeneous integral equation plus an infinite set of nonhomogeneous integral equations. A simplifying feature is that each of the nonhomogeneous integral equations which determines the $f^{(n)}$, $n \geq 1$, has the identical associated homogeneous integral equation, $J(f^{(0)},f^{(n)}) = 0$.

Boltzmann's H theorem allows us to immediately write the unique solution to Equation (6–8), which has the form of a local Maxwellian distribution function, that is,

$$f^{(0)}(\mathbf{q},\mathbf{v},t) = \frac{\rho^{(0)} e^{-(\mathbf{v}-\mathbf{u}^{(0)})^2/2RT^{(0)}}}{(2\pi RT^{(0)})^{3/2}} \qquad (6\text{–}13)$$

where

$$\begin{bmatrix} \rho^{(0)} \\ \rho^{(0)}\mathbf{u}^{(0)} \\ 3\rho^{(0)}RT^{(0)} \end{bmatrix} = \int d\mathbf{v}\, f^{(0)}(\mathbf{q},\mathbf{v},t) \begin{bmatrix} 1 \\ \mathbf{v} \\ (\mathbf{v} - \mathbf{u}^{(0)})^2 \end{bmatrix} \qquad (6\text{–}14)$$

It is important to note that $f^{(0)}$ is not the local Maxwellian corresponding to the thermo-fluid state of the system since $\rho^{(0)}$, $\mathbf{u}^{(0)}$, $T^{(0)}$ are not necessarily equal to ρ, $\mathbf{u}$, T. The former set of variables are found by taking moments with respect to $f^{(0)}$, while the latter set are determined by the moments of f, and the two sets of moments are not necessarily equal.

Proceeding to determine the other $f^{(n)}$, we first write $f^{(n)} = f^{(0)}g^{(n)}$ so that the homogeneous equation satisfied by $g^{(n)}$ is of the familiar form

$$J(f^{(0)},f^{(0)}g^{(n)}) = f^{(0)}L(g^{(n)}) = 0 \tag{6–15}$$

Therefore, $g^{(n)}$ is a sum of the eigenfunctions of L having zero for an eigenvalue. These eigenfunctions are just the summational invariants, so that the solution to Equation (6–15) is

$$g^{(n)} = \sum_{i=1}^{5} \gamma_i^{(n)}(\mathbf{q},t)\hat{\psi}_i(\mathbf{v}) \tag{6–16}$$

where the $\hat{\psi}_i$, $(i = 1 - 5)$, are the five summational invariants 1, v_α, v^2. In Chapter 4 we showed that Equation (6–15) was a homogeneous Fredholm equation of the second kind, so that we can make use of the properties of such equations, and their associated inhomogeneous equations, which we listed in the preceding section. Let us first consider the equation for $f^{(1)}$, Equation (6–9). Making use of the Fredholm theorem [Equation (6–4)] we see that solutions of the equation for $f^{(1)}$ exist only if

$$\int d\mathbf{v}\, Df^{(0)}\hat{\psi}_i = 0 \qquad i = 1 - 5 \tag{6–17}$$

The set of equations obtained from Equation (6–17) are just the usual macroscopic thermo-fluid equations with the substitution of $\rho^{(0)}$, $\mathbf{u}^{(0)}$, $T^{(0)}$ for ρ, $\mathbf{u}$, T, and with $\mathbf{Q} = 0$, $P_{\alpha\beta} = \rho^{(0)}RT^{(0)}\delta_{\alpha\beta} \equiv p^{(0)}\delta_{\alpha\beta}$.

A formal solution for $f^{(n)}$ can also be obtained. We denote the particular solution of Equation (6–12) as $\bar{f}^{(n)}$, and in terms of this quantity we have

$$f^{(n)} = \bar{f}^{(n)} + \sum_{i=1}^{5} \gamma_i^{(n)}f^{(0)}\hat{\psi}_i \tag{6–18}$$

The constants $\gamma_i^{(n)}$ are determined by applying the Fredholm theorem to the equation for $f^{(n+1)}$, which requires that they satisfy the relationship

$$\int d\mathbf{v}\left[Df^{(n)} - \sum_{m=1}^{n} J(f^{(m)},f^{(n+1-m)})\right]\hat{\psi}_i = 0 \qquad i = 1 - 5$$

The second term can be shown to be identically zero by making use of the symmetry properties of $J(\ \ ,\ \ )$, and we thus arrive at

$$\int d\mathbf{v}\, Df^{(n)}\hat{\psi}_i = 0 \qquad i = 1 - 5 \tag{6–19}$$

as the condition the $\gamma_i^{(n)}$ must satisfy. (The $\bar{f}^{(n)}$ which will appear in the above equation when Equation (6–18) is used for $f^{(n)}$ can be presumed known in what follows.) It will be convenient to rewrite the above equation in the following form:

$$\int d\mathbf{v}\, D\bar{f}^{(n)}\hat{\psi}_i + \sum_{j=1}^{5} \int d\mathbf{v}\, D(f^{(0)}\gamma_j^{(n)})\hat{\psi}_i\hat{\psi}_j = 0 \tag{6–20}$$

or, more concisely, as

$$\frac{\partial}{\partial t} a_{i\beta}\gamma_\beta^{(n)} + \frac{\partial}{\partial q_\alpha}(b_{i\beta}^{\alpha}\gamma_\beta^{(n)} + c_{i\alpha}^{(n)}) = 0 \tag{6–21}$$

with the definitions

$$\begin{bmatrix} a_{ij} \\ b_{ij}^{\alpha} \end{bmatrix} = \int d\mathbf{v}\, f^{(0)}\hat{\psi}_i\hat{\psi}_j \begin{bmatrix} 1 \\ v_\alpha \end{bmatrix}$$
$$c_{ij}^{(n)} = \int d\mathbf{v}\, \bar{f}^{(n)}\hat{\psi}_i v_j \tag{6–22}$$

The set of differential equations specified by Equation (6–21) serve to uniquely specify the coefficients $\gamma_i^{(n)}(\mathbf{q},t)$, which then determine $f^{(n)}$, provided only that proper initial data is furnished. Let us now introduce the functions

$$\begin{bmatrix} \rho_i^{(n)} \\ \rho_i \end{bmatrix} = \int d\mathbf{v}\, \hat{\psi}_i \begin{bmatrix} f^{(n)} \\ f \end{bmatrix} \tag{6–23}$$

so that the ρ_i are given by the sum over the $\rho_i^{(n)}$,

$$\rho_i = \sum_n \delta^n \rho_i^{(n)}$$

Specifically, we have

$$\begin{aligned} \rho_1 &= \rho \\ \rho_{2,3,4} &= \rho u_{x,y,z} \\ \rho_5 &= 2\rho(e + \tfrac{1}{2}u^2) \end{aligned}$$

Once again, invoking the second Fredholm theorem, we require

$$\int d\mathbf{v}\, \hat{\psi}_i \bar{f}^{(n)} = 0$$

so that the $\rho_i^{(n)}$ can be written explicitly in terms of a_{ij} and $\gamma_i^{(n)}$ as

$$\begin{aligned} \rho_i^{(n)} &= \sum_j \int d\mathbf{v}\, \hat{\psi}_i\hat{\psi}_j\gamma_i^{(n)}f^{(0)} \\ &= a_{i\beta}\gamma_\beta^{(n)} \end{aligned} \tag{6–24}$$

The above relationship can be used to determine $f^{(n)}$ in terms of the initial values of $\rho_i^{(n)}, \rho_i^{(n-1)}, \cdots, \rho_i^{(0)}$ by substituting for the γ_β in Equation (6–21).

Hilbert's result now follows from the following considerations. We have

$$\rho_i(\mathbf{q},0;\delta) = \sum_n \delta^n \rho_i^{(n)}(\mathbf{q},0) \tag{6–25}$$

which, with the above prescription, provides the normal solution to Equation (6–6). If we consider instead the equation

$$D\tilde{f} = \frac{1}{\mu} J(\tilde{f}) \tag{6–6$'$}$$

a solution can clearly be found by retracing the above steps. Further, if we now require

$$\begin{aligned} \int d\mathbf{v}\, \hat{\psi}_i \tilde{f}^{(0)} &= \rho_i(\mathbf{q},0;\delta) \\ \int d\mathbf{v}\, \hat{\psi}_i \tilde{f}^{(n)} &= 0 \qquad n \geq 1 \end{aligned} \tag{6–26}$$

then $\tilde{f} = \tilde{f}(\mathbf{q},\mathbf{v},t;\delta,\mu)$. Since the function $\tilde{f}(\mathbf{q},\mathbf{v},t;\delta,\delta)$ is a solution to Equation (6–6′) for fixed δ with initial conditions given by Equation (6–26), it then follows that $\tilde{f}(\mathbf{q},\mathbf{v},t;\delta,\delta)$ depends only on the initial values of its first five moments. If we now use Equation (6–25) to expand $\rho_i(\mathbf{q},0;\delta)$ we see that $\tilde{f}(\mathbf{q},\mathbf{v},t;\delta,\delta)$ has the same initial values as $f(\mathbf{q},\mathbf{v},t;\delta)$, and therefore we conclude that these functions are identical, and that f can itself be determined uniquely in terms of the initial values of its first five moments. But the choice $t = 0$ is completely arbitrary, so that if $f(\mathbf{q},\mathbf{v},0)$ is determined by $\rho_i(\mathbf{q},0)$, then we can also state that $f(\mathbf{q},\mathbf{v},t)$ is determined by $\rho_i(\mathbf{q},t)$, which is Hilbert's theorem.

6–3 THE CHAPMAN–ENSKOG PROCEDURE

In a series of papers published between 1911 and 1917, Sydney Chapman (in England) and David Enskog (in Sweden) independently described how somewhat more general solutions to Boltzmann's equation than those of the Hilbert type could be obtained. The plausibility of their method is greatly enhanced, however, by the Hilbert theorem. The primary use which has been made of the Chapman–Enskog procedure is the calculation of transport coefficients (remember that in the macroscopic theories these quantities are unknown!). This method of treating the Boltzmann equation is not particularly well-suited for producing solutions to specific boundary or initial value problems or as the basis for the treatment of generalized flows, and we shall therefore defer the treatment of these topics to later chapters.

The Chapman–Enskog procedure has two distinguishing features. The first lies in considering a direct expansion of the equations for the moments of f (in the last section we referred to the first five moments of f

as ρ_i, $i = 1, 2, \cdots, 5$, and we shall continue this notational shorthand here) rather than expanding the ρ_i themselves. The second feature of the Chapman–Enskog procedure lies in the introduction of the *ansatz* that the entire time dependence of f is solely through ρ, $\mathbf{u}$, T; that is, instead of treating f as a function of t it is treated as a functional of ρ, $\mathbf{u}$, T. Clearly, in light of Hilbert's theorem, this *ansatz* appears reasonable. In fact, the *ansatz* would be rigorous if f were expanded in a power series in the small parameter δ [see Equation (6–6) and the discussion immediately preceding it]. However, in the Chapman–Enskog scheme we look for solutions of the form

$$f = \sum_{n=0}^{\infty} \delta^n f^{(n)} \tag{6–27}$$

where

$$\begin{bmatrix} \rho \\ \rho\mathbf{u} \\ 3\rho RT \end{bmatrix} = \int d\mathbf{v}\, f^{(0)} \begin{bmatrix} 1 \\ \mathbf{v} \\ (\mathbf{v}-\mathbf{u})^2 \end{bmatrix} \tag{6–28}$$

$$0 = \int d\mathbf{v}\, f^{(n)} \begin{bmatrix} 1 \\ \mathbf{v} \\ (\mathbf{v}-\mathbf{u})^2 \end{bmatrix} \qquad n \geq 1 \tag{6–29}$$

and, as a consequence of imposing the conditions stipulated by Equations (6–28) and (6–29), the $f^{(n)}$ are themselves functions of δ. Formally, the Chapman–Enskog *ansatz* leads to the replacement

$$f(\mathbf{q},\mathbf{v},t) \rightarrow f(\mathbf{q},\mathbf{v};\rho,\mathbf{u},T) \tag{6–30}$$

which allows the following substitution to be made on the left-hand side of the Boltzmann equation:

$$\frac{\partial f}{\partial t} = \frac{\partial f}{\partial \rho}\frac{\partial \rho}{\partial t} + \frac{\partial f}{\partial \mathbf{u}} \cdot \frac{\partial \mathbf{u}}{\partial t} + \frac{\partial f}{\partial T}\frac{\partial T}{\partial t} \tag{6–31}$$

Although the ρ_i are not expanded in δ, their derivatives are, since these quantities are given in terms of P and $\mathbf{Q}$, which as higher moments of f must be expanded following Equation (6–27). Thus we must write

$$\begin{aligned} \frac{\partial \rho}{\partial t} &= -\nabla \cdot \rho\mathbf{u} \\ \frac{\partial \mathbf{u}}{\partial t} &= -(\mathbf{u}\cdot\nabla)\mathbf{u} - \frac{1}{\rho}\nabla \cdot \sum_{n=0}^{\infty} \delta^n \mathsf{P}^{(n)} \\ \frac{\partial T}{\partial t} &= -\mathbf{u}\cdot\nabla T - \frac{2}{3R\rho}\left[\sum_{n=0}^{\infty} \delta^n [\mathsf{P}^{(n)} : \nabla\mathbf{u}] + \nabla \cdot \sum_{n=0}^{\infty} \delta^n \mathbf{Q}^{(n)}\right] \end{aligned} \tag{6–32}$$

where

$$\begin{bmatrix} \mathsf{P}^{(n)} \\ \mathbf{Q}^{(n)} \end{bmatrix} = \int d\mathbf{v}\, f^{(n)} \begin{bmatrix} \mathbf{v}_0\mathbf{v}_0 \\ \mathbf{v}_0 v_0^2 \end{bmatrix} \tag{6–33}$$

The equation for $(\partial T/\partial t)$ is derived similarly to those for $(\partial \rho/\partial t)$ and $(\partial \mathbf{u}/\partial T)$ (see Problem 2–8).

In order to facilitate the proper ordering of terms in what follows we now introduce the operator $(\partial_n/\partial t)$ which picks out the $O(\delta^n)$ term of the time derivative of the operand, so that we have, for example,

$$\begin{aligned} \frac{\partial_0 \rho}{\partial t} &= -\nabla \cdot \rho\mathbf{u} \\ \frac{\partial_n \rho}{\partial t} &= 0 \qquad n \geq 1 \\ \frac{\partial_0 \mathbf{u}}{\partial t} &= -(\mathbf{u} \cdot \nabla)\mathbf{u} - \frac{1}{\rho}\nabla \cdot \mathsf{P}^{(0)} \\ \frac{\partial_n \mathbf{u}}{\partial t} &= -\frac{1}{\rho}\nabla \cdot \mathsf{P}^{(n)} \qquad n \geq 1 \end{aligned} \tag{6–34}$$

and so forth.

The substitution for $(\partial f/\partial t)$ can now be completed by combining Equations (6–27), (6–31), and (6–34) to write

$$\begin{aligned} \frac{\partial f}{\partial t} &= \sum_{n=0}^{\infty} \delta^n \sum_{m=0}^{n} \frac{\partial_m}{\partial t}\left(\rho \frac{\partial}{\partial \rho} + \mathbf{u} \cdot \frac{\partial}{\partial \mathbf{u}} + T \frac{\partial}{\partial T}\right) f^{(n-m)} \\ &\equiv \sum_{n=0}^{\infty} \delta^n \frac{\partial_n f}{\partial t} \end{aligned} \tag{6–35}$$

The remaining terms in Boltzmann's equation are expanded in a straight forward manner using Equation (6–27) directly. The left-hand side of the equation is then written as

$$\frac{\partial f}{\partial t} + \mathbf{v} \cdot \frac{\partial f}{\partial \mathbf{q}} = \sum_{n=0}^{\infty} \delta^n \frac{D_n f}{Dt} \equiv \sum_{n=0}^{\infty} \delta^n \left[\frac{\partial_n f}{\partial t} + \mathbf{v} \cdot \frac{\partial f^{(n)}}{\partial \mathbf{q}}\right] \tag{6–36}$$

while the right-side becomes

$$\frac{1}{\delta} J(f) = \frac{1}{\delta} J(f^{(0)}) + \sum_{n=1}^{\infty} \sum_{m=0}^{n} \delta^{n-1} J(f^{(m)}, f^{(n-m)}) \tag{6–37}$$

[see Equation (6–11) for the definition of $J(f^{(m)}, f^{(n)})$].

Equating the terms of like order in the small parameter δ in Equations (6–36) and (6–37) then gives a set of equations which determines the $f^{(n)}$; that is, considering terms of $O(1/\delta)$ we find

$$J(f^{(0)}) = 0 \tag{6–38}$$

and considering terms of $O(\delta^0)$ gives an equation for $f^{(1)}$, and so forth.

The above considerations have led us to a set of integral equations for the $f^{(n)}$, just as in Hilbert's treatment. At each step in the development the unknown, $f^{(n)}$, appears in the term $J(f^{(0)},f^{(n)})$. This term can be related to the linearized collision operator L; specifically, we replace $f^{(n)}$ by $f^{(0)}\phi^{(n)}$ thereby obtaining

$$J(f^{(0)},f^{(n)}) = J(f^{(0)},f^{(0)}\phi^{(n)}) = f^{(0)}L(\phi^{(n)}) \tag{6–39}$$

where $\phi^{(n)}$ is now the unknown. (This follows directly from Equations (6–28) and (6–38), which imply that $f^{(0)}$ is the local Maxwellian distribution function.) Thus, the solubility conditions for the equation which determines $f^{(n)}$ are the Fredholm conditions

$$\int d\mathbf{v}\,\hat{\psi}_i\,\frac{D_0 f}{Dt} = 0 \qquad n = 1$$

$$\int d\mathbf{v}\,\hat{\psi}_i\left\{\frac{D_{n-1}f}{Dt} - \sum_{m=1}^{n-1} J(f^{(m)},f^{(n-m)})\right\} = 0 \qquad n > 1 \tag{6–40}$$

or

$$\int d\mathbf{v}\,\hat{\psi}_i\,\frac{D_{n-1}f}{Dt} = 0 \qquad n > 1 \tag{6–41}$$

since from the symmetry properties of J it follows that

$$\int d\mathbf{v}\,J(g,h)\hat{\psi}_i = 0$$

The set of equations (6–41) are identical to the thermo-fluid equations obtained from setting

$$f = \sum_{i=1}^{n-1} \delta^i f^{(i)}$$

that is, they are the same as Equation (6–32) with terms up to $O(\delta^{n-1})$ retained. Thus the solubility conditions can be considered to be satisfied for all the $f^{(n)}$.

The solution for f which is obtained by taking the first n terms in the δ expansion is called the nth order Chapman–Enskog solution. Only the first- and second-order Chapman–Enskog solutions will be considered here, since it is at present unclear whether the higher-order solutions are generally physically relevant. (For certain problems the third-order

Chapman–Enskog solution has been employed with successful results. The resulting thermo-fluid equations are called the Burnett equations, after the English mathematician David Burnett, who first carried out the third-order theory in 1935.)

6–4 THE FIRST-ORDER CHAPMAN–ENSKOG SOLUTION

Although we have already anticipated the results of the first-order solution in the preceding section, it seems worthwhile to formally carry through the necessary steps leading to these results as a prelude to considering the second-order solution. The equation for $f^{(0)}$ is obtained by equating terms of $O(1/\delta)$ in Equations (6–36) and (6–37), and doing this, we obtain Equation (6–38). This equation is to be solved subject to the conditions imposed by Equation (6–28), which states that the first five moments of $f^{(0)}$ are the thermo-fluid variables ρ, $\mathbf{u}$, T. Boltzmann's H theorem then tells us that the unique solution to these equations is the local Maxwellian distribution. Consequently we have $\mathbf{Q}^{(0)} = 0$, and $\mathsf{P}_{\alpha\beta}{}^{(0)} = p\delta_{\alpha\beta} = \rho RT\delta_{\alpha\beta}$ and we see that in the first-order approximation the thermo-fluid equations correspond to the usual equations for an inviscid fluid. These are called the Euhler equations. With $f^{(0)}$ determined we can now proceed to the second-order solution.

6–5 THE SECOND-ORDER CHAPMAN–ENSKOG SOLUTION

The equation for $f^{(1)}$ is obtained by equating the terms of $O(\delta^0)$ which appear in Equations (6–36) and (6–37). Doing this we find

$$\frac{D_0 f}{Dt} = J(f^{(0)},f^{(1)}) + J(f^{(1)},f^{(0)})$$
$$= \frac{1}{m}\int d\mathbf{v}_2\, d\theta\, d\epsilon\, B(\theta,V)[f_1{}^{(0)\prime}f_2{}^{(1)\prime} + f_2{}^{(0)\prime}f_1{}^{(1)\prime} - f_1{}^{(0)}f_2{}^{(1)} - f_2{}^{(0)}f_1{}^{(1)}] \tag{6–42}$$

The solution of the above equation, as we will see shortly, is truly a gruesome task. With this warning we now continue.

The inhomogeneous term appearing in Equation (6–42), D_0f/Dt, can be made explicit since $f^{(0)}$ is now known. Writing out this term, we find, after some straightforward manipulation,

$$\frac{D_0 f}{Dt} = f^{(0)}\left\{\left(\frac{v_0{}^2}{2RT} - \frac{5}{2}\right)v_{0\alpha}\frac{\partial \ln T}{\partial q_\alpha} + \frac{1}{RT}\left(v_{0\alpha}v_{0\beta} - \frac{v_0{}^2}{3}\delta_{\alpha\beta}\right)\frac{\partial u_\alpha}{\partial q_\beta}\right\} \tag{6–43}$$

The unknown, $f^{(1)}$, appears in the integral term on the right side of Equation (6–42). Introducing the function $\Phi = f^{(1)}/f^{(0)}$ that term can be written as

$$\frac{1}{m}\int d\mathbf{v}_2\, d\theta\, d\epsilon\, B(\theta,V) f_1{}^{(0)} f_2{}^{(0)} [\Phi_1' + \Phi_2' - \Phi_1 - \Phi_2]$$
$$= f_1{}^{(0)} L(\Phi) \equiv \rho^2 I(\Phi) \qquad (6\text{–}44)$$

where we have made use of the fact that

$$f_1{}^{(0)\prime} f_2{}^{(0)\prime} = f_1{}^{(0)} f_2{}^{(0)}$$

(The operator I is introduced only as a convenience to bring our notation more closely in accord with that of Chapman and Cowling's book, *The Mathematical Theory of Non-Uniform Gases*, the standard reference for the material contained in this section.) Finally, if we introduce the dimensionless variables $\mathbf{w} = \mathbf{v}/(2RT)^{1/2}$, and $\mathbf{w}_0 = \mathbf{v}_0/(2RT)^{1/2}$ we obtain the following equation for Φ:

$$f_0(\mathbf{v}) \left[(w_0{}^2 - \tfrac{5}{2}) v_{0\alpha} \frac{\partial \ln T}{\partial q_\alpha} + 2 \left(w_{0\alpha} w_{0\beta} - \frac{w_0{}^2}{3} \delta_{\alpha\beta} \right) \frac{\partial u_\beta}{\partial q_\alpha} \right] = \rho^2 I(\Phi) \qquad (6\text{–}45)$$

which we must then solve for Φ.

The following comments can be made regarding the structure of the solution of Equation (6–45) following a cursory inspection of that equation.

1. We already know the solution to the associated homogeneous equation; it is just

$$\sum_{i=1}^{5} a_i(\mathbf{q},t) \hat{\psi}_i$$

2. Since f, and therefore Φ, must be a scalar quantity, the solution of Equation (6–45) must be a linear combination of a scalar product of some vector and $(\partial \ln T/\partial \mathbf{q})$, and the scalar product of some tensor and $(\partial \mathbf{u}/\partial \mathbf{q})$.

We can therefore write

$$\Phi(\mathbf{v}) = \frac{1}{\rho}(2RT)^{1/2}\mathbf{A} \cdot \frac{\partial \ln T}{\partial \mathbf{q}} + \frac{1}{\rho}\mathbf{B} : \frac{\partial \mathbf{u}}{\partial \mathbf{q}} + a_1 + \mathbf{a} \cdot \mathbf{v}_0 + a_5 v_0{}^2 \qquad (6\text{–}46)$$

Substituting the above expression back into Equation (6–45) and equating coefficients of $(\partial \ln T/\partial \mathbf{q})$, $(\partial \mathbf{u}/\partial \mathbf{q})$ we obtain the following equations for the unknown functions $\mathbf{A}$ and $\mathbf{B}$:

$$\rho I(\mathbf{A}) = f^{(0)}(w_0{}^2 - \tfrac{5}{2})\mathbf{w}_0 \qquad (6\text{–}47)$$

$$\rho I(B_{\alpha\beta}) = 2f^{(0)} \left(w_{0\alpha} w_{0\beta} - \frac{w_0{}^2}{3} \delta_{\alpha\beta} \right) \equiv 2f^{(0)} (\mathbf{w}_0{}^0\mathbf{w}_0)_{\alpha\beta} \qquad (6\text{–}48)$$

Since $f^{(0)}$ is linear in ρ we can conclude that $\mathbf{A}$ and $\mathbf{B}$ will be independent of this quantity, a fact which we shall recall later.

The inhomogeneous terms in Equations (6–47) and (6–48) are composed of the quantities ρ, T, $\mathbf{w}_0$, and therefore $\mathbf{A}$ and $\mathbf{B}$ must be completely describable in ρ, T, $\mathbf{w}_0$. The only vector which can be formed from this set is $\mathbf{w}_0$, and therefore

$$\mathbf{A} = A(T,w_0)\mathbf{w}_0 \tag{6–49}$$

The form of $\mathbf{B}$ can be deduced by similar considerations. First we note that

$$\rho I(B_{\alpha\alpha}) = 0$$

and second that

$$\rho I(B_{\alpha\beta} - B_{\beta\alpha}) = 0$$

so that $\mathbf{B}$ is nondivergent ($B_{\alpha\alpha} = 0$) and symmetric ($B_{\alpha\beta} = B_{\beta\alpha}$). Therefore

$$\mathbf{B} = B(T,w_0)\mathbf{w}_0{}^0\mathbf{w}_0 \tag{6–50}$$

since $\mathbf{w}_0{}^0\mathbf{w}_0$ is the only symmetrical nondivergent tensor which can be formed from T and $\mathbf{w}_0$.

To complete our specification of the form of Φ we now show that the a's which appear in Equation (6–46) can be set equal to zero, so that the homogeneous solution does not directly contribute to Φ. The second Fredholm condition states that Φ will be a unique solution to Equation (6–45) provided that

$$\int d\mathbf{v}\, \hat{\psi}_i f^{(0)}\Phi = 0 \tag{6–51}$$

Collecting our above results, and substituting into the above equation, we find

$$\int d\mathbf{v}\, \hat{\psi}_i f^{(0)} \left\{ \frac{1}{\rho} (2RT)^{1/2} A(w_0)\mathbf{w}_0 \cdot \frac{\partial \ln T}{\partial \mathbf{q}} + \frac{1}{\rho} B(w_0)\mathbf{w}_0{}^0\mathbf{w}_0 : \frac{\partial \mathbf{u}}{\partial \mathbf{q}} + a_1 + \mathbf{a}\cdot\mathbf{v}_0 + a_5 v_0{}^2 \right\} = 0 \tag{6–52}$$

(We will no longer show the T dependence of $\mathbf{A}$ and $\mathbf{B}$.) After some simple algebraic manipulation the above equations can be written as

$$\int d\mathbf{v}\, f^{(0)}(a_1 + a_5 v_0{}^2) = 0$$

$$\int d\mathbf{v}\, f^{(0)} \left(A(w_0) \frac{\partial \ln T}{\partial \mathbf{q}} + \mathbf{a} \right) v_0{}^2 = 0$$

$$\int d\mathbf{v}\, f^{(0)}(a_1 + a_5 v_0{}^2) v_0{}^2 = 0 \tag{6–53}$$

From the first and last of the above equations we see that $a_1 = a_5 = 0$. The middle equation indicates that $\mathbf{a}$ is proportional to $(\partial \ln T/\partial \mathbf{q})$, and we can absorb this term into the term including $\mathbf{A}$ by imposing the following condition:

$$\int d\mathbf{v}\, f^{(0)} A(w_0) w_0^2 = 0 \tag{6-54}$$

The above considerations have led us to the following expression for Φ:

$$\Phi = \frac{1}{\rho}(2RT)^{1/2}\mathbf{A}\cdot\frac{\partial \ln T}{\partial \mathbf{q}} + \frac{1}{\rho}\mathsf{B} : \frac{\partial \mathbf{u}}{\partial \mathbf{q}} \tag{6-55}$$

The second Chapman–Enskog approximation is then found by solving the integral equations (6–47) and (6–48) for $\mathbf{A}$ and B subject to the condition of Equation (6–54). In practice, as we will see in the following sections, these quantities are not directly evaluated, since only certain integrals involving them, which turn out to be simpler to calculate than $\mathbf{A}$ and B themselves, are of primary interest.

6–6 EXPRESSIONS FOR THE THERMAL CONDUCTIVITY AND VISCOSITY

In the second Chapman–Enskog approximation the heat flux is

$$\begin{aligned}\mathbf{Q}^{(1)} &= \frac{1}{2}\int d\mathbf{v}\ \mathbf{v}_0 v_0^2 f^{(0)}\Phi \\ &= \frac{1}{2\rho}\int d\mathbf{v}\ \mathbf{v}_0 v_0^2 (2RT)^{1/2}\mathbf{A}\cdot\frac{\partial \ln T}{\partial \mathbf{q}}\end{aligned} \tag{6-56}$$

The contribution to $\mathbf{Q}^{(1)}$ of the component of Φ containing B is identically zero, as seen by a simple consideration of parity. The above integral can be written in a more convenient form by noting that if $F(\mathbf{y})$ is any scalar function of the vector $\mathbf{y}$, and of odd degree in at least one component of $\mathbf{y}$, then if $\mathbf{z}$ is any other vector,

$$\begin{aligned}\int d\mathbf{y}\, F(\mathbf{y})\mathbf{y}(\mathbf{z}\cdot\mathbf{y}) &= \mathbf{z}\cdot\int d\mathbf{y}\, F(\mathbf{y})\mathbf{y}\mathbf{y} \\ &= \tfrac{1}{3}\mathbf{z}\cdot\mathsf{U}\int d\mathbf{y}\, F(\mathbf{y})y^2 \\ &= \tfrac{1}{3}\mathbf{z}\int d\mathbf{y}\, F(\mathbf{y})y^2\end{aligned} \tag{6-57}$$

The above result can be used to rewrite Equation (6–56) as

$$\begin{aligned}\mathbf{Q}^{(1)} &= \left\{\frac{2}{3\rho}(RT)^2\int d\mathbf{v}\, f^{(0)} w_0^4 A(w_0)\right\}\frac{\partial \ln T}{\partial \mathbf{q}} \\ &= \left\{\frac{2}{3}\frac{R^2T}{\rho}\int d\mathbf{v}\, f^{(0)}(w_0^4 - \tfrac{5}{2}w_0^2)A(w_0)\right\}\frac{\partial T}{\partial \mathbf{q}} \\ &= \left\{\frac{2}{3}\frac{R^2T}{\rho}\int d\mathbf{v}\, f^{(0)}A(w_0)\mathbf{w}_0(w_0^2 - \tfrac{5}{2})\cdot\mathbf{w}_0\right\}\frac{\partial T}{\partial \mathbf{q}}\end{aligned} \tag{6-58}$$

(we have also made use of Equation (6–54) in arriving at the above expression). The term $(w_0^2 - \frac{5}{2})\mathbf{w}_0$, which appears in the above integrand, is equal [see Equation (6–47)] to $\rho I(\mathbf{A})$, so that we can also write

$$\mathbf{Q}^{(1)} = \frac{2R^2T}{3} \int d\mathbf{v}\, \mathbf{A} \cdot I(\mathbf{A}) \frac{\partial T}{\partial \mathbf{q}}$$

$$\equiv \frac{-2R^2T}{3} [\mathbf{A},\mathbf{A}] \frac{\partial T}{\partial \mathbf{q}} \equiv -\kappa \frac{\partial T}{\partial \mathbf{q}} \tag{6–59}$$

Thus the second Chapman–Enskog approximation leads us to the Fourier law of heat conduction. The coefficient κ, which relates the heat flow to the temperature gradient, is called the coefficient of thermal conductivity. As we have mentioned earlier, one of the advantages of the microscopic theory is that the possibility exists, entirely within the framework of the theory, of explicitly determining the transport coefficients. This is, in fact, what we have just laid the groundwork for doing for the thermal conductivity. In the following section we will show how κ may be explicitly calculated.

There is still additional information to be extracted from the second Chapman–Enskog approximation, for we have yet to consider $\mathsf{P}^{(1)}$. Substituting for Φ, and again making use of a simple parity argument, we find

$$\mathsf{P}^{(1)} = \int d\mathbf{v}\, f^{(0)}\Phi \mathbf{v}_0\mathbf{v}_0$$

$$= \frac{2RT}{\rho} \int d\mathbf{v}\, f^{(0)} B(w_0) \mathbf{w}_0{}^0\mathbf{w}_0 : \frac{\partial \mathbf{u}}{\partial \mathbf{q}}\, \mathbf{w}_0\mathbf{w}_0$$

$$= \frac{2RT}{5\rho} \int d\mathbf{v}\, f^{(0)} B(w_0)(\mathbf{w}_0{}^0\mathbf{w}_0 : \mathbf{w}_0\mathbf{w}_0) \left[\mathsf{D} - \frac{D_{\alpha\alpha}}{3}\mathsf{U}\right] \tag{6–60}$$

where the rate of strain tensor D is defined as

$$D_{\alpha\beta} = \frac{1}{2}\left[\frac{\partial u_\alpha}{\partial q_\beta} + \frac{\partial u_\beta}{\partial q_\alpha}\right] = D_{\beta\alpha} \tag{6–61}$$

The final form of Equation (6–60) has been obtained through the following considerations. Let $\mathbf{y}$ again be an arbitrary vector, $F(\mathbf{y})$ be an arbitrary scalar function, and C be an arbitrary tensor. Then

$$\mathbf{y}^0\mathbf{y} : \mathsf{C} = \mathbf{y}\mathbf{y} : \overset{0}{\mathsf{C}}$$
$$\hat{\mathbf{y}\mathbf{y}} : \mathsf{C} = \mathbf{y}\mathbf{y} : \hat{\mathsf{C}} \tag{6–62}$$

and we can write

$$\int d\mathbf{y}\, F(\mathbf{y})(\mathbf{y}^0\mathbf{y} : \mathsf{C})\mathbf{y}\mathbf{y} = \int d\mathbf{y}\, F(\mathbf{y})(\mathbf{y}\hat{^0}\mathbf{y}) : \mathsf{C})\mathbf{y}\mathbf{y} = \int d\mathbf{y}\, F(\mathbf{y})(\mathbf{y}\mathbf{y} : \overset{0}{\hat{\mathsf{C}}})\mathbf{y}\mathbf{y} \tag{6–63}$$

(where $\hat{C}_{\alpha\beta} = \frac{1}{2}(C_{\alpha\beta} + C_{\beta\alpha})$. The above integral is a tensor having diagonal elements, for example,

$$\int d\mathbf{y}\, F(\mathbf{y}) y_i^2 (y_i^2 \overset{0}{\hat{C}}_{ii} + y_j^2 \overset{0}{\hat{C}}_{jj} + y_k^2 \overset{0}{\hat{C}}_{kk}) \tag{6-64}$$

If the $\mathbf{y}$ integration is carried out in a polar coordinate system we find

$$\begin{aligned} \int d\mathbf{y}\, F(\mathbf{y}) y_i^2 y_j^2 &= \frac{1}{5} \int d\mathbf{y}\, F(\mathbf{y}) y^4 \qquad i = j \\ &= \frac{1}{15} \int d\mathbf{y}\, F(\mathbf{y}) y^4 \qquad i \neq j \end{aligned} \tag{6-65}$$

so that, with Equation (6-64) we have

$$\int d\mathbf{y}\, F(\mathbf{y}) y_i^2 (\mathbf{yy} : \overset{0}{\mathbf{C}}) = \tfrac{2}{15} \overset{0}{\hat{C}}_{ii} \int d\mathbf{y}\, F(\mathbf{y}) y^4 \tag{6-66}$$

The off-diagonal elements in Equation (6-63) are, for example,

$$\begin{aligned} \int d\mathbf{y}\, F(\mathbf{y}) y_i y_j (\mathbf{yy} : \overset{0}{\mathbf{C}}) &= 2 \int d\mathbf{y}\, F(\mathbf{y}) y_i^2 y_j^2 \overset{0}{\hat{C}}_{ij} \\ &= \tfrac{2}{15} \overset{0}{\hat{C}}_{ij} \int d\mathbf{y}\, F(\mathbf{y}) y^4 \end{aligned} \tag{6-67}$$

where we have made use of Equation (6-66). Combining the above results, we find

$$\int d\mathbf{y}\, F(\mathbf{y}) (\mathbf{y}^0\mathbf{y} : \mathbf{C}) \mathbf{yy} = \tfrac{2}{15} \overset{0}{\hat{\mathbf{C}}} \int d\mathbf{y}\, F(\mathbf{y}) y^4 \tag{6-68}$$

One last identity is required to establish the desired result, namely,

$$\begin{aligned} \mathbf{y}^0\mathbf{y} : \mathbf{y}^0\mathbf{y} &= \mathbf{y}^0\mathbf{y} : [\mathbf{yy} - \tfrac{1}{3}\mathbf{U}(\mathbf{U} : \mathbf{yy})] \\ &= \mathbf{y}^0\mathbf{y} : \mathbf{yy} \\ &= \mathbf{yy} : \mathbf{yy} - \tfrac{1}{3} y^2 (\mathbf{U} : \mathbf{yy}) \\ &= (\mathbf{y} \cdot \mathbf{y})^2 - \tfrac{1}{3}\mathbf{y}^4 = \tfrac{2}{3}\mathbf{y}^4 \end{aligned} \tag{6-69}$$

so that we may write, finally,

$$\int d\mathbf{y}\, F(\mathbf{y}) (\mathbf{y}^0\mathbf{y} : \mathbf{C}) = \tfrac{1}{5} \overset{0}{\hat{\mathbf{C}}} \int d\mathbf{y}\, F(\mathbf{y}) (\mathbf{y}^0\mathbf{y} : \mathbf{y}^0\mathbf{y}) \tag{6-70}$$

which indicates how we arrived at the last line of Equation (6-60).

Returning now to Equation (6-60), we rewrite this equation as

$$\begin{aligned} \mathbf{P}^{(1)} &= \left(\frac{2RT}{5\rho} \int d\mathbf{v}\, f^{(0)} \mathbf{B} : \mathbf{w}_0\mathbf{w}_0 \right) (\mathbf{D} - \tfrac{1}{3} D_{\alpha\alpha} \mathbf{U}) \\ &= \left(\frac{RT}{5} \int d\mathbf{v}\, \mathbf{B} : I(\mathbf{B}) \right) (\mathbf{D} - \tfrac{1}{3} D_{\alpha\alpha} \mathbf{U}) \\ &= -\frac{RT}{5} [\mathbf{B}, \mathbf{B}] (\mathbf{D} - \tfrac{1}{3} D_{\alpha\alpha} \mathbf{U}) \equiv -2\mu (\mathbf{D} - \tfrac{1}{3} D_{\alpha\alpha} \mathbf{U}) \end{aligned} \tag{6-71}$$

In the second Chapman–Enskog approximation we then have

$$\mathsf{P} = \mathsf{P}^{(0)} + \mathsf{P}^{(1)} = p\mathsf{U} - 2\mu(\mathsf{D} - \tfrac{1}{3}D_{\alpha\alpha}\mathsf{U}) \tag{6–72}$$

which is the usual macroscopic form for the pressure tensor given by the Newton–Stokes relationship. Again, the pertinent transport coefficient, in this case the coefficient of viscosity, is given explicitly.

At this point we have, in principle, achieved one of our major objectives. The macroscopic laws of heat conduction and stress have been derived with explicit expressions for the transport coefficients κ and μ. We see that these coefficients will depend on the microscopic properties of the system under consideration, since the bracketed expressions directly involve the collision integral, which depends in an explicit manner on the interparticle forces.

In order to evaluate κ and μ we do not have to explicitly compute **A** and B, but only the bracket expressions [**A**,**A**] and [B,B] which are considerably easier to calculate. Even the calculation of the bracket quantities is, in general, quite tedious, and we do not feel that it is inconsistent with our purpose here, which is to emphasize the post-Chapman–Cowling developments in kinetic theory, to present only a skeletal outline of the procedure. Full details may be found in Chapman and Cowling's book.

6–7 CALCULATION OF THE TRANSPORT COEFFICIENTS

The procedure used in calculating κ and μ is essentially the same for both quantities. Chapman and Enskog's original procedure was to expand **A** and B in terms of polynomials in $\mathbf{w}_0$; in 1935 Burnett showed that it is particularly convenient to choose Sonnine polynomials for this expansion. The reason for the particular suitability of these functions became clear only later when it was shown that for Maxwell molecules they are the eigenfunctions of the linearized Boltzmann collision operator, which appears as the integral operator in the equations determining **A** and B. (This was first pointed out in an unpublished report by (Madame) Wang Chang and Uhlenbeck in 1952, and Uhlenbeck has since given the credit to Wang Chang. The identical result was also, independently, proved by Waldmann in his *Handbuch* article in 1958.) Thus **A** and B can be found exactly for Maxwell molecules; they are directly related to the eigenvalues of L. For other interparticle potentials **A** and B are found approximately by what is essentially a perturbation about the Maxwell molecule result.

The Sonnine polynomial of order n, index m, and argument x^2 is defined as

$$S_m^{(n)}(x^2) = \sum_{l=0}^{n} \frac{\Gamma(n+m+1)(-x^2)^l}{l!(n-1)!\Gamma(l+m+1)} \tag{6–73}$$

where $\Gamma(n)$ is the standard gamma function,

$$\Gamma(n) = (n-1)!$$

$$\Gamma(n+\tfrac{1}{2}) = 1\cdot 3\cdot 5 \cdots (2n-3)(2n-1)\frac{\sqrt{\pi}}{2^n}$$

for n any integer greater than zero. The $S_m^{(n)}$ satisfy the following orthogonality conditions:

$$\int_0^\infty dx\, x^{2m+2}e^{-x^2}S_{m+1/2}^{(n)}(x^2)S_{m+1/2}^{(n')}(x^2) = \frac{1}{2n!}\Gamma(m+n+\tfrac{3}{2})\delta_{nn'} \tag{6–74}$$

Let us consider the case of arbitrary interparticle potential, and treat Maxwell molecules as a special case. Expanding **A** and **B** in the appropriate Sonnine polynomials, we write

$$\begin{aligned}\mathbf{A} &= A\mathbf{w}_0 = \sum_{n=0} a_n S_{3/2}^{(n)}(w_0^2)\mathbf{w}_0 \equiv \sum_{n=0} a_n \mathbf{a}^{(n)}\\ \mathsf{B} &= B\mathbf{w}_0{}^0\mathbf{w}_0 = \sum_{n=0} b_n S_{5/2}^{(n)}(w_0^2)\mathbf{w}_0{}^0\mathbf{w}_0 \equiv \sum_{n=0} b_n \mathbf{b}^{(n)}\end{aligned} \tag{6–75}$$

with a_n, b_n coefficients to be determined. Consider the evaluation of **A**. As we have formulated the problem, this quantity must satisfy Equation (6–54), which we write below for convenience:

$$\int d\mathbf{v}\, f^{(0)}A(w_0)w_0^2 = 0$$

The above condition is satisfied by setting $a_0 = 0$. To show this we rewrite Equation (6–54) in the following form:

$$\begin{aligned}\int d\mathbf{w}_0\, e^{-w_0^2}w_0^2 \sum_{n=0} a_n S_{3/2}^{(n)}(w_0^2) &= 4\pi \sum_{n=0} a_n \int_0^\infty dw_0\, e^{-w_0^2}S_{3/2}^{(0)}(w_0^2)S_{3/2}^{(n)}(w_0^2)w_0^4\\ &= 2\pi a_0\Gamma(\tfrac{5}{2}) = 0\end{aligned}$$

from which it follows that $a_0 = 0$.

The remaining coefficients are found through the following considerations. Define the quantities

$$a_{nm} = [\mathbf{a}^{(n)},\mathbf{a}^{(m)}] \tag{6-76}$$

$$\begin{aligned}\alpha_n &= \frac{1}{\rho}\int d\mathbf{v}\, f^{(0)}(w_0^2 - \tfrac{5}{2})\mathbf{w}_0 \cdot \mathbf{a}^{(n)} \\ &= \int d\mathbf{v}\, I(\mathbf{A}) \cdot \mathbf{a}^{(n)} = -[\mathbf{a}^{(n)},\mathbf{A}] \\ &\qquad\qquad = -\sum_{m=1} a_m[\mathbf{a}^{(n)},\mathbf{a}^{(m)}] \\ &\qquad\qquad = -\sum_{m=1} a_m a_{nm}\end{aligned} \tag{6-77}$$

The a_m are thus determined by solving an infinite set of linear inhomogeneous equations. The solution is given formally through the application of Cramer's rule:

$$a_m = \frac{\mathcal{A}_m}{\mathcal{A}} \tag{6-78}$$

Here $\mathcal{A}$ is the determinant formed by the a_{nm}, and $\mathcal{A}_m$ is the determinant obtained from $\mathcal{A}$ by substituting $\alpha_1, \alpha_2, \cdots$ for the mth column.

In general, both $\mathcal{A}_m$ and $\mathcal{A}$ will diverge, so that in practice the a_m cannot be directly evaluated using Equation (6–78). In order to obtain approximate results for the a_m, a complicated truncation procedure has to be introduced for the determinants. For Maxwell molecules, however, an exact result can be obtained by noting that since $\mathbf{w}_0(w_0^2 - \frac{5}{2})$ is an eigenfunction of L, $a_m = 0$, $m \geq 2$. For this case we then find

$$[\mathbf{A},\mathbf{A}] = \frac{\alpha_1^2}{a_{11}} \tag{6-79}$$

The above expression also gives the first approximation for non-Maxwell molecules. (Note that a_{11} itself will depend on the interparticle potential.) In general, the correction terms are small. For a power law potential they go as $(s - 5)^2/4(s - 1)(11s - 13)$.

The calculation of α_1 is quite straightforward. In general, we will have

$$\begin{aligned}\alpha_m &= \frac{1}{\rho}\int d\mathbf{w}_0\, f^{(0)}(w_0^2 - \tfrac{5}{2})w_0^2 S_{3/2}^{(m)}(w_0^2) \\ &= -2\pi^{-1/2}\int_0^\infty d(w_0^2)e^{-w_0^2}w_0^3 S_{3/2}^{(1)}(w_0^2)S_{3/2}^{(m)}(w_0^2) \\ &= -\tfrac{15}{4}\delta_{m1}\end{aligned} \tag{6-80}$$

The a_{11}, which depend on the form of the interparticle potential, are, by comparison, quite tedious to compute. The result, for power law potentials, is

$$a_{11} = 4(\pi RT)^{1/2}\left(\frac{mK}{2RT}\right)^{2/(s-1)} A_2(s)\Gamma\left(4 - \frac{2}{s-1}\right) \qquad (6\text{–}81)$$

The values of the function $A_2(s)$ are computed and listed in Chapman and Cowling. Of particular interest are Maxwell molecules, $A_2(5) = 0.436$, Lennard–Jones repulsive potential, $A_2(7) = 0.357$, and hard spheres, $A_2(\infty) = 0.333$.

For any interparticle potential the thermal conductivity (and the viscosity) is independent of density, and depends on temperature as $T^{(1/2)+[2/(s-1)]}$, so that for Maxwell molecules κ is directly proportional to temperature. The demonstration that κ is independent of density in a rarefied gas confirms a result obtained by Maxwell by considerably less general methods, and is one of the major results of the kinetic theory.

The viscosity coefficient μ may be determined by following the same procedure used in obtaining κ. For Maxwell molecules the result is

$$\mu = \kappa \frac{4}{15R} \qquad (6\text{–}82)$$

and for general power law potentials the above result is a valid first approximation. We will consider the determination of μ in some detail in the Problem Section.

6–8 VALIDITY OF THE CHAPMAN–ENSKOG RESULTS

In this section we briefly take up the question of when the Chapman–Enskog solution can be expected to be valid. In order to make some rough estimates let us consider hard sphere molecules, for which $a_{11} \propto \sigma^2 T^{-1/2}$ so that $\Phi \propto 1/\rho\sigma^2$. For the approximation we are considering to be valid, Φ must be small (compared to one) and consequently $(1/\rho\sigma^2)(1/T)(\partial T/\partial \mathbf{q})$ must be small. This condition will be satisfied if $\Delta T/T$ is itself small, where ΔT is the temperature change per mean free path. Thus the Chapman–Enskog approximation can be expected to be good in regimes where the system is not too far from equilibrium, in the sense described above. A similar restriction can be shown to hold for the velocity gradients.

It is important to note another criteria that must also be taken into account in assessing the validity of the Chapman–Enskog solution for a particular situation. This is that the density cannot be too small since

$\Phi \propto \rho^{-1}$. Thus, although we expect Boltzmann's equation to be particularly good at low densities, there is a point beyond which the system becomes so rarefied that the collision mechanism is ineffective in establishing a local Maxwellian distribution, and $f^{(0)}$ is no longer a good zeroth order approximation to f. In fact, we have already seen in Section 6–2 that we cannot expect normal solutions to be valid in extremely rarefied gases so the above conclusions should not come as too big of a surprise to us here.

This concludes our discussion of the calculation of transport coefficients (with the exception of some special topics which will be dealt with later), which in today's fast moving scientific milieu comprises the "ancient history" of kinetic theory. We now can move on to the more recent applications of Boltzmann's equation.

References

As a reference to Section 6–1, see any book on the theory of integral equations, for example:

1. F. Smithies, *Integral Equations*. Cambridge, England: Cambridge University Press, 1962.

The Hilbert and Chapman–Enskog techniques were first proposed by:

2. D. Hilbert, *Math. Ann.*, vol. 72, p. 562, 1912.
3. S. Chapman, *Phil. Trans. Royal Soc.*, ser. A, vol. 216, p. 279, 1916.
4. ——, *Phil. Trans. Royal Soc.*, ser. A, vol. 217, p. 115, 1917.
5. D. Enskog, Doctoral dissertation, Uppsala University, Uppsala, Sweden, 1917.

The original Chapman–Enskog procedure was simplified considerably through the use of Sonnine polynomials by:

6. D. Burnett, *Proc. London Math. Soc.*, vol. 39, p. 385, 1935.
7. ——, *Proc. London Math. Soc.*, vol. 40, p. 382, 1935.

Copious results for the transport coefficients for various molecular interactions can be found in:

8. S. Chapman and T. G. Cowling (1952). See General References.

9. J. O. Hirschfelder, C. F. Curtiss, and R. B. Bird, *Molecular Theory of Gases and Liquids*. New York: Wiley, 1954.

10. L. Waldmann (1958). See General References.

The relationship between Hilbert and Chapman–Enskog solutions to actual solutions of Boltzmann's equation has been discussed by:

11. H. Grad, *Phys. Fluids*, vol. 6, p. 147, 1963.

12. H. P. McKean, Jr., *J. Math. Phys.*, vol. 8, p. 547, 1967.

Problems

6–1. Verify the result given in Equation (6–43) for (D_0f/Dt).

6–2. Define quantities β_n, b_{nm} analogous to the α_n, a_{nm} defined in Equation (6–76), and show that for Maxwell molecules $[\mathbf{B},\mathbf{B}] = \beta_1{}^2/b_{11}$.

6–3. Explicitly evaluate the β_n defined in Problem 6–2, and use this result, together with the result of Problem 6–2, to write μ for Maxwell molecules in terms of b_{11}.

6–4. According to the laws of thermodynamics, when an ideal monatomic gas undergoes some quasistatic adiabatic process, the temperature and density are related according to $\rho \propto T^{3/2}$. Show that the thermo-fluid equations in the first Chapman–Enskog approximation also satisfy this relationship, and explain why.

6–5. Prove that $\int d\mathbf{v}\, J(g,h)\hat{\psi}_i(\mathbf{v}) = 0$, where the $\hat{\psi}_i(\mathbf{v})$ are the summational invariants.

6–6. A Boltzmann equation with $J(f)$ replaced by $g^*J(f)$ appears in the theory of dense gases, with $g^* = g^*(\mathbf{q},t)$. What are the values of κ and μ which are found using this equation (in terms of the values calculated using Boltzmann's equation)?

6–7. For a flow with no heat flow, prove that the form of the distribution function obtained by minimizing H subject to the constraints

$$\int d\mathbf{v}\, f\begin{pmatrix} 1 \\ \mathbf{v} \\ \mathbf{v}_0\mathbf{v}_0 \end{pmatrix} = \begin{pmatrix} \rho \\ \rho\mathbf{u} \\ \mathsf{P} \end{pmatrix}$$

is equal in first order to the corresponding second Chapman–Enskog approximation.

6–8. Use the results of the second Chapman–Enskog approximation to obtain the closed set of equations for ρ, $\mathbf{u}$, T in this approximation.

6–9. Use the results of the second Chapman–Enskog approximation to evaluate $(D_1 f/Dt)$, and explain what this could imply with regard to the structure of $f^{(2)}$. Why doesn't the above result allow a definite statement to be made regarding the structure of $f^{(2)}$?

CHAPTER 7

moment method solutions

7-1 INTRODUCTION

Following the work of Chapman and Enskog there was a period of almost thirty years during which the main preoccupation in kinetic theory was the computation of transport coefficients for various intermolecular potentials following the procedures of Chapman and Enskog. The basic problem, of directly solving the Boltzmann equation for f rather than determining transport coefficients, does not seem to have been considered during this period. Also left unanswered were important questions pertaining to the properties of nonnormal solutions of Boltzmann's equation. Clearly there are important regimes where the normal solutions cannot be expected to be valid, for example, in a strong shock wave or a boundary layer; not surprisingly these are also the regimes for which the conventional macroscopic thermo-fluid description fails. Strangely enough, the first attack on these problems was by Maxwell, and considerably pre-dates the work of Chapman and Enskog.

Maxwell's approach to solving Boltzmann's equation makes strong use of the properties of Maxwell molecules (which, of course, explains the terminology), and therefore lacks the generality necessary for a completely satisfactory theory. It was not until 1949, with the publication of Grad's dissertation, that a general theory of nonnormal solutions of Boltzmann's equation became available. (It is interesting to note that the main results of Chapman and Enskog also comprised their doctoral dissertations.) The publication of Grad's dissertation opened a new era in kinetic theory, concerned with the explicit solution of Boltzmann's equation for specific flow problems, and in the ensuing twenty years a great deal of work has been done in this direction (there still remains much to do, however).

Grad's method is very simple in conception, but fairly complicated in the working out of the attendant details. This method is based on an expansion of f in a set of n-dimensional tensors which are the Cartesian equivalent of Sonnine polynomials. These are the n-dimensional Hermite tensor polynomials, which are a generalization of the standard Hermite polynomials used so often in mathematical physics. In the next section we describe these functions, develop the notation necessary to carry through Grad's method, and list results for later reference. Then we briefly consider Maxwell's theory, which leads us directly to its generalization and extension by the method of Grad, which we develop and discuss in the concluding sections of this chapter.

7-2 MULTIDIMENSIONAL HERMITE TENSOR POLYNOMIALS

The n-dimensional Hermite tensor polynomials developed by Grad are generalizations of the ordinary Hermite polynomials which are standard fare in mathematical physics. The reader familiar with these functions will recall that the Hermite polynomial of index n and argument x is given by the formula

$$H_n(x) = (-)^n e^{x^2} \frac{d^n}{dx^n} e^{-x^2} \tag{7-1}$$

where x is a scalar. The quantities used in Grad's theory are the tensor equivalent of the $H_n(x)$; specifically, we define the rank n tensor polynomial $H_{i_1,i_2,\ldots,i_n}{}^{(n)}(\mathbf{x}) \equiv H_i{}^{(n)}(\mathbf{x})$ through the formula

$$\begin{aligned} H_i{}^{(n)}(\mathbf{x}) &= (-)^n e^{\mathbf{x}^2/2} \left(\frac{\partial}{\partial x_{i_1}} \frac{\partial}{\partial x_{i_2}} \cdots \frac{\partial}{\partial x_{i_n}}\right) e^{-\mathbf{x}^2/2} \\ &= (-)^n e^{\mathbf{x}^2/2} \boldsymbol{\nabla}^{(n)} e^{-\mathbf{x}^2/2} \end{aligned} \tag{7-2}$$

where $\boldsymbol{\nabla}^{(n)}$ is a generalized gradient operator. The $H_i{}^{(n)}$ can also be written in terms of the function $\omega(x) = (2\pi)^{-3/2} e^{-x^2/2}$ which will prove useful in Section 7-4.

$$H_i(\mathbf{x}) = (-)^n \omega^{-1}(x) \boldsymbol{\nabla}^{(n)} \omega(x) \tag{7-3}$$

The importance of the Hermite tensor polynomials in kinetic theory stems from the fact that they are eigenfunctions of the linearized collision operator L for Maxwell molecules.

The first few tensor polynomials can be easily computed from Equation (7-2) or (7-3), and we find

$$\begin{aligned} H^{(0)} &= 1 \\ H_i{}^{(1)} &= x_i \\ H_{ij}{}^{(2)} &= x_i x_j - \delta_{ij} \\ H_{ijk}{}^{(3)} &= x_i x_j x_k - (x_i \delta_{jk} + x_j \delta_{ik} + x_k \delta_{ij}) \end{aligned} \tag{7-4}$$

The tensor polynomials are orthonormal with respect to the weight ω,

$$\int d\mathbf{x}\, \omega(x) H_i^{(n)}(\mathbf{x}) H_j^{(n)}(\mathbf{x}) = \delta_{ij}^{(n)} \tag{7-5}$$

where $\delta_{ij}^{(n)}$ is zero unless the subscripts $i_1, i_2, \cdots, i_n$ are a permutation of the subscripts $j_1, j_2, \cdots, j_n$.

7-3 MAXWELL'S MOMENT METHOD

In the case of Maxwell molecules it is possible to obtain a great deal of information pertaining to the thermo-fluid properties without attempting to directly solve Boltzmann's equation. The method by which this is done is due to Maxwell and, since it can be considered as the fore-runner of the more general method of Grad, it seems worthwhile to review it prior to discussing Grad's work.

To begin we rewrite the Boltzmann equation in terms of the Eulerian (or substantive) derivative

$$\frac{D}{Dt} \equiv \frac{\partial}{\partial t} + \mathbf{u} \cdot \frac{\partial}{\partial \mathbf{q}}$$

in the following form:

$$\frac{Df}{Dt} + \mathbf{v}_0 \cdot \frac{\partial f}{\partial \mathbf{q}} - \frac{D\mathbf{u}}{Dt} \cdot \frac{\partial f}{\partial \mathbf{v}_0} - \frac{\partial f}{\partial \mathbf{v}_0}\mathbf{v} : \frac{\partial \mathbf{u}}{\partial \mathbf{q}} = J(f) \tag{7-6}$$

The Maxwell transport equation for the molecular property $\pi(\mathbf{v})$ is obtained from Boltzmann's equation by multiplying it by π and integrating, so that we find

$$\begin{aligned}\frac{D}{Dt}(\rho\pi_{\text{av}}) &+ \rho\pi_{\text{av}} \frac{\partial}{\partial \mathbf{q}} \cdot \mathbf{u} + \frac{\partial}{\partial \mathbf{q}} \cdot \rho(\pi\mathbf{v}_0)_{\text{av}} \\ &- \rho\left(\frac{D\pi}{Dt} + \mathbf{v}_0 \cdot \frac{\partial \pi}{\partial \mathbf{q}}\right)_{\text{av}} - \frac{D\mathbf{u}}{Dt} \cdot \left(\frac{\partial \pi}{\partial \mathbf{v}_0}\right)_{\text{av}} - \left(\frac{\partial \pi}{\partial \mathbf{v}_0}\mathbf{v}_0\right)_{\text{av}} : \frac{\partial \mathbf{u}}{\alpha \mathbf{q}} \\ &= \frac{1}{m} \int d\mathbf{v}_1\, d\mathbf{v}_2\, d\theta\, d\epsilon\, B(\theta,V)\pi(f_1'f_2' - f_1f_2) \equiv \rho(\Delta\pi)_{\text{av}}\end{aligned} \tag{7-7}$$

where

$$\int d\mathbf{v}\, \pi f = \rho\pi_{\text{av}}$$

Setting $\pi = 1$, $\mathbf{v}_0$ we again obtain the familiar thermo-fluid equations of motion. In general, if $\pi = \pi(\mathbf{v}_0)$, we will have

$$\begin{aligned}D(\pi)_{\text{av}} &\equiv \frac{D}{Dt}(\rho\pi_{\text{av}}) + \rho\pi_{\text{av}} \frac{\partial}{\partial \mathbf{q}} \cdot \mathbf{u} - \frac{\partial}{\partial \mathbf{q}} \cdot \mathsf{P} \cdot \left(\frac{\partial \pi}{\partial \mathbf{v}_0}\right)_{\text{av}} + \rho \frac{\partial}{\partial \mathbf{v}_0}(\pi\mathbf{v}_0) : \frac{\partial \mathbf{u}}{\partial \mathbf{q}} \\ &= \rho(\Delta\pi)_{\text{av}}\end{aligned} \tag{7-8}$$

As a particular example of Equation (7–8) let us consider $\pi = \mathbf{v}_0\mathbf{v}_0$. After some algebraic manipulation, we obtain

$$\frac{D}{Dt}\mathsf{P} + \mathsf{P}\frac{\partial}{\partial \mathbf{q}}\cdot\mathbf{u} + \left(\mathsf{P}\cdot\frac{\partial}{\partial \mathbf{q}}\right)\mathbf{u} + \left(\mathsf{P}\cdot\frac{\partial}{\partial \mathbf{q}}\mathbf{u}\right)^{\dagger} + \frac{\partial}{\partial \mathbf{q}}\mathfrak{Q} = \rho(\Delta\mathbf{v}_0\,\mathbf{v}_0)_{\mathrm{av}} \tag{7–9}$$

where

$$\mathfrak{Q} = \int d\mathbf{v}_0\,\mathbf{v}_0\mathbf{v}_0\mathbf{v}_0 f$$

and

$$C_{ij}^{\dagger} = C_{ji}$$

Continuing, we can obtain an infinite set of equations for the moments of f. Since f is uniquely determined by the set of all its moments, we could, in principle, obtain the same information contained in the solution to Boltzmann's equation by solving the infinite set of moment equations. Let us denote the set of moments of f taken with respect to $\mathbf{v}_0$ by $M_n{}^0$,

$$M_n{}^0 = \int d\mathbf{v}_0\, f\mathbf{v}_0{}^n \tag{7–10}$$

The set of equations for the $M_n{}^0$ is obtained from Equation (7–7) by choosing $\pi = \mathbf{v}_0{}^n$; note that this will not be a closed set of equations. The contribution of the collision term does not introduce any higher moments (remember we are considering Maxwell molecules; the preceding statement is not true for general interparticle potentials). In fact, the contribution of the collision term can be expressed in terms of products of $M_k{}^0$, $k \leq n$. The $(\partial/\partial\mathbf{q})\cdot\rho(\pi\mathbf{v}_0)_{\mathrm{av}}$ term, however, will introduce a term $(\partial/\partial\mathbf{q})\rho M_{n+1}{}^0$ into the equation for $M_n{}^0$. Thus the equation for $M_0{}^0$ contains $M_1{}^0$, the equation for $M_1{}^0$ contains $M_2{}^0$, and so forth. In order to close these equations it is clear that some approximation scheme involving a truncation will have to be used.

An iterative type of approximation scheme to effect solutions for the set of equations for the $M_n{}^0$ was devised by Maxwell, and proceeds as follows. Let the moments in the mth-order solution be denoted by a superscript (m), for example $\mathsf{P}^{(m)}$, $\mathfrak{Q}^{(m)}$, $\cdots$. An iteration is defined by setting, for arbitrary m,

$$\begin{aligned} \Delta^{(0)}(\mathbf{v}_0{}^n)_{\mathrm{av}} &= 0 \\ D^{(m)}(\mathbf{v}_0{}^n)_{\mathrm{av}} &= \Delta^{(m+1)}(\mathbf{v}_0{}^n)_{\mathrm{av}} \qquad m \geq 0 \end{aligned} \tag{7–11}$$

which defines an infinite set of determined equations (for $n = 1, 2, \cdots$) in each order (given m). In lowest order we set $m = 0$ and find

$$\Delta^{(0)}(\mathbf{v}_0{}^n)_{\mathrm{av}} = 0 \tag{7–12}$$

This implies that f is a local Maxwellian, so that, for example,

$$P_{ij}^{(0)} = p\delta_{ij} = \rho RT\delta_{ij} \tag{7-13}$$
$$Q_i^{(0)} = 0 \tag{7-14}$$

In the next approximation the following results are of particular interest:

$$(P_{ij}^{(1)} - p\delta_{ij}) = -2\mu\,(\boldsymbol{D} - \tfrac{1}{3}D_{\alpha\alpha}\boldsymbol{U}) \tag{7-15}$$
$$\mathbf{Q}^{(1)} = -\kappa\frac{\partial T}{\partial \mathbf{q}} \tag{7-16}$$

where μ and κ are the identical values found by Chapman and Enskog for Maxwell molecules. The thermo-fluid equations in this approximation are the Navier–Stokes equations, so that in second order, the Maxwell moment method is fully compatible with the corresponding Chapman–Enskog approximation. This compatibility is not maintained in higher-order approximations, and the question arises as to which method is to be preferred. First we note that the Chapman–Enskog method is applicable, at the expense of a great deal of computational labor (much of which can now be done by computers), to general interparticle force laws, whereas Maxwell's method is restricted to Maxwell molecules. On the other hand, we must view this positive aspect of the Chapman–Enskog method as actually being far outweighed by the major negative aspect of the method, namely the restriction of its validity to normal solutions, which limits the physical situations in which the method can be applied.

What would be most desirable, of course, would be a method of solving Boltzmann's equation which includes the strong points of both the Maxwell and Chapman–Enskog methods, so that, if necessary, non-normal solutions for general interparticle potentials could be obtained. These are just the features enjoyed by Grad's method of solution for Boltzmann's equation, and we now turn our attention to its development.

7–4 GRAD'S MOMENT METHOD

In this section we will closely follow Grad's original exposition of the development of the general moment method of solution for Boltzmann's equation. We therefore begin by introducing with Grad a dimensionless velocity variable $\hat{\mathbf{v}}_0$ and a dimensionless distribution function $\hat{f}$, which are defined as

$$\mathbf{v}_0 = \hat{\mathbf{v}}_0(RT)^{1/2} \tag{7-17}$$
$$f = \rho(RT)^{-3/2}\hat{f} \tag{7-18}$$

so that

$$\int d\hat{\mathbf{v}}_0 \hat{f} \begin{bmatrix} 1 \\ \hat{\mathbf{v}}_0 \\ \hat{v}_0^2 \end{bmatrix} = \begin{bmatrix} 1 \\ 0 \\ 3 \end{bmatrix} \tag{7-19}$$

Next, we expand f in the Hermite tensor functions discussed in Section 7–2 about a local Maxwellian distribution, so that

$$f = f_{\mathrm{LM}} \sum_{n=0} \frac{1}{n!} a_\alpha^{(n)}(\mathbf{q},t) H_\alpha^{(n)}(\hat{\mathbf{v}}_0) \tag{7-20}$$

or

$$\hat{f} = \hat{f}_{\mathrm{LM}} \sum_{n=0} \frac{1}{n!} a_\alpha^{(n)}(\mathbf{q},t) H_\alpha^{(n)}(\hat{\mathbf{v}}_0) \tag{7-21}$$

where $\hat{f}_{\mathrm{LM}}$, the local Maxwellian in dimensionless units, is identical to $\omega(\hat{v}_0)$. The expansion coefficients $a_i^{(n)}(\mathbf{q},t)$ are, as shown, functions of space and time. Multiplying Equation (7–21) (or (7–20)) by $H_j^{(n)}$ and integrating over $\hat{\mathbf{v}}_0$ serves to determine these coefficients in terms of $\hat{f}$ (or f). We find, using the orthogonality relationship of Equation (7–5),

$$\begin{aligned} a_i^{(n)} &= \textstyle\int d\hat{\mathbf{v}}_0 \hat{f} H_i^{(n)} \\ &= \rho^{-1} \textstyle\int d\mathbf{v}_0 f H_i^{(n)} \end{aligned} \tag{7-22}$$

Specifically, we will have

$$\begin{aligned} a^{(0)} &= 1 \\ a_i^{(1)} &= 0 \\ a_{ij}^{(2)} &= \frac{P_{ij} - p\delta_{ij}}{p} \equiv \frac{p_{ij}}{p} \\ a_{ijk}^{(3)} &= \frac{Q_{ijk}}{p(RT)^{1/2}} \end{aligned} \tag{7-23}$$

In general, it is found that any of the coefficients $a_i^{(n)}$ can be expressed as a linear combination of the M_k^0, $k \leq n$.

An equation for $\hat{f}$ can be obtained by substituting Equation (7–18) into Boltzmann's equation, so that we have

$$\frac{\partial \hat{f}_1}{\partial t} + \mathbf{v}_1 \cdot \frac{\partial \hat{f}_1}{\partial \mathbf{q}} + \hat{f}_1 \left(\frac{\mathbf{p}}{p} : \frac{\partial \mathbf{u}}{\partial \mathbf{q}} + \frac{1}{2p} \frac{\partial}{\partial \mathbf{q}} \cdot \mathbf{Q} \right) + \mathbf{v}_0 \cdot \hat{f}_1 \frac{\partial}{\partial \mathbf{q}} \left(\ln \frac{\rho}{(RT)^{3/2}} \right)$$
$$= \frac{\rho}{m} \int d\hat{\mathbf{v}}_{0_2}\, d\theta\, d\epsilon\, B(\theta, \hat{V}(RT)^{1/2})(\hat{f}_1'\hat{f}_2' - \hat{f}_1\hat{f}_2) \tag{7-24}$$

where we have made use of the conservation equations to simplify the last two terms on the left-hand side. (In keeping with our present notation, we have also written $\mathbf{V} = \hat{\mathbf{V}}(RT)^{1/2}$.) The above equation can now be used to generate a set of equations for the $a_i^{(n)}$. This is done by multiplying Equation (7–24) by $H_i^{(n)}$ and integrating over $\mathbf{v}_0$, which gives

$$\frac{D}{Dt}a^{(n)} + \frac{\partial \mathbf{u}}{\partial q_\alpha}a_\alpha^{(n)} + \frac{n}{2RT}\left(\frac{DRT}{Dt}\right)a^{(n)}$$
$$+ (RT)^{1/2}\frac{\partial a_\alpha^{(n+1)}}{\partial q_\alpha} + (RT)^{1/2}a_\alpha^{(n+1)}\frac{\partial}{\partial q_\alpha}\ln\left(\rho(RT)^{(n+1)/2}\right)$$
$$+ \left((RT)^{-1/2}\frac{D\mathbf{u}}{Dt} + (RT)^{1/2}\frac{\partial}{\partial \mathbf{q}_s}\ln\rho(RT)^{(n+1)/2}\right)a^{(n-1)}$$
$$+ \left((RT)^{-1/2}\frac{\partial RT}{\partial q_\alpha}\boldsymbol{\delta}a_\alpha^{(n-1)}\right) + \left((RT)^{-1}\frac{DRT}{Dt}\boldsymbol{\delta} + \frac{\partial \mathbf{u}}{\partial \mathbf{q}_s}\right)a^{(n-2)}$$
$$+ (RT)^{-1/2}\frac{\partial RT}{\partial \mathbf{q}_s}\boldsymbol{\delta}a^{(n-3)} = J^{(n)} \qquad (7\text{–}25)$$

In writing the above equation we have employed the following notational shorthand. The operator $\partial/\partial \mathbf{q}_s$, represents a symmetrized gradient, so that, for example,

$$\left(\frac{\partial a^{(1)}}{\partial \mathbf{q}_s}\right)_{ij} = \frac{\partial a_j^{(1)}}{\partial q_i} + \frac{\partial a_i^{(1)}}{\partial q_j}$$
$$\left(\frac{\partial a^{(2)}}{\partial \mathbf{q}_s}\right)_{ijk} = \frac{\partial a_{jk}^{(2)}}{\partial q_i} + \frac{\partial a_{ik}^{(2)}}{\partial q_j} + \frac{\partial a_{ij}^{(2)}}{\partial q_k} \qquad (7\text{–}26)$$

The Kronecker delta tensor $\boldsymbol{\delta}$ is just δ_{ij}, whereas δ_i, which we will use later, is the sum of all terms in which an i is affixed to $\boldsymbol{\delta}$. The collision term $J^{(n)}$ is

$$J^{(n)} = \frac{\rho}{m}\int d\hat{\mathbf{v}}_{0_1}\,d\hat{\mathbf{v}}_{0_2}\,d\theta\,d\epsilon\,B(\theta,\hat{V}(RT)^{1/2})\cdot\{\hat{f}_1'\hat{f}_2' - \hat{f}_1\hat{f}_2\}H^{(n)} \qquad (7\text{–}27)$$

which can be written, making use of the results of Section 4–1, as

$$J^{(n)} = J_\nu^{(n)} = \frac{\rho}{2m}\int d\hat{\mathbf{v}}_{0_1}\,d\hat{\mathbf{v}}_{0_2}\,\hat{f}(\mathbf{v}_{0_1})\,\hat{f}(\mathbf{v}_{0_2})I_\nu^{(n)} \qquad (7\text{–}28)$$

where

$$I_\nu^{(n)} = \int d\theta\,d\epsilon\,B(\theta,\hat{V}(RT)^{1/2})[H_\nu^{(n)}] \qquad (7\text{–}29)$$
$$[H_\nu^{(n)}] = H_\nu^{(n)}(\hat{\mathbf{v}}_{0_1}') + H_\nu^{(n)}(\hat{\mathbf{v}}_{0_2}') - H_\nu^{(n)}(\hat{\mathbf{v}}_{0_1}) - H_\nu^{(n)}(\hat{\mathbf{v}}_{0_2}) \qquad (7\text{–}30)$$

We can write $J^{(n)}$ as an infinite quadratic sum of coefficients $a_i^{(n)}$ by replacing the $\hat{f}$s which appear in the expression for this quantity by their expansion in the Hermite functions. Thus, we can also write

$$J_\nu^{(n)} = \frac{\rho}{2m}\sum_{r,s=0}\beta_{\nu\rho\sigma}^{(nrs)}a_\rho^{(r)}a_\sigma^{(s)} \qquad (7\text{–}31)$$

where

$$\beta_{\nu\rho\sigma}{}^{(nrs)} = \frac{1}{r!s!}\int d\hat{\mathbf{v}}_{0_1}\, d\hat{\mathbf{v}}_{0_2}\, \hat{f}_{LM_1}\hat{f}_{LM_2}H_\rho{}^{(r)}(\hat{\mathbf{v}}_{0_1})H_\sigma{}^{(s)}(\hat{\mathbf{v}}_{0_2})I_\nu{}^{(n)}$$

$$= \frac{1}{r!s!}\int d\hat{\mathbf{v}}_{0_1}\, d\hat{\mathbf{v}}_{0_2}\, d\theta\, d\epsilon\, B(\theta,\hat{V}(RT)^{1/2}) \cdot \hat{f}_{LM_1}\hat{f}_{LM_2}H_\rho{}^{(r)}(\hat{\mathbf{v}}_{0_1})H_\sigma{}^{(s)}(\hat{\mathbf{v}}_{0_2})[H_\nu{}^{(n)}]$$

$$= \frac{1}{r!s!}\int d\hat{\mathbf{v}}_{0_1}\, d\hat{\mathbf{v}}_{0_2}\, d\theta\, d\epsilon\, B(\theta,\hat{V}(RT)^{1/2})\hat{f}_{LM_1}\hat{f}_{LM_2}H_\nu{}^{(n)}(\hat{\mathbf{v}}_{0_1}) \cdot [H_\rho{}^{(r)}H_\sigma{}^{(s)}] \qquad (7\text{–}32)$$

In the case of Maxwell molecules the β's will be pure numbers, independent of temperature, since $B(\theta,V)$ is independent of the velocity variable $\mathbf{v}$. Further, because of the orthogonality properties of the Hermite functions, β will vanish unless $n = r + s$, so that in this case $J^{(n)}$ contains a finite quadratic sum of coefficients of order not exceeding n. For general interparticle potentials, however, $J^{(n)}$ includes all the coefficients $a^{(k)}$, $k = 1, 2, \cdots, n, n+1, \cdots$.

The discussion indicates that there are two distinct problems which will confront us in attempting to solve the equations for the $a^{(n)}$ [Equation (7–25)]. First, the left-hand side of this set of equations always contains a term with $a^{(n+1)}$, and second, for non-Maxwell molecules the right-hand side, $J^{(n)}$ contains all the $a^{(k)}$. The procedure which we will adopt in effecting a closure of these equations is as follows. The expansion for f will be truncated after a given number of terms. If we retain n terms in the expansion, the $J^{(n)}$ is automatically truncated at $r = n$, $s = n$ and no higher-order coefficients will appear in the expression for this term. For Maxwell molecules this truncation is unnecessary, due to the orthogonality properties of the $H^{(n)}$. To complete the closure we drop the $a^{(n+1)}$ term which appears on the left-hand side of the equation for $a^{(n)}$. We then have, for given n, a set of closed equations for $a^{(k)}$, $k = 0, 1, 2, \cdots, n$.

The first significant generalization of the Navier–Stokes theory is obtained when we truncate the expansion for f at $n = 3$, so that

$$f = f_{LM}\left(1 + \frac{a_\alpha{}^{(2)}H_\alpha{}^{(2)}}{2} + \frac{a_\alpha{}^{(3)}H_\alpha{}^{(3)}}{6}\right) \qquad (7\text{–}33)$$

or, using the explicit form of the $H_{ijk}{}^{(n)}$,

$$f = f_{LM}\left(1 + p_{\alpha\beta}\frac{v_{0\alpha}v_{0\beta}}{2pRT} + \frac{Q_{\alpha\beta\partial}}{6p(RT)^2}v_{0\alpha}v_{0\beta}v_{0\partial} - \frac{Q_\alpha v_{0\alpha}}{2pRT}\right) \qquad (7\text{–}34)$$

Here we see quite explicitly the dependence of f on its moments ρ, $\mathbf{u}$, T, p_{ij}, Q_{ijk} in contradistinction to the Chapman–Enskog form of f which depends only on ρ, $\mathbf{u}$, T. In what follows we will not consider the dependence of f on the full $\mathcal{Q}_{ijk}$, but only on that part of this moment which is related to the heat flux vector $Q_i = (\mathcal{Q}_{i\alpha\alpha}/2)$. Thus, in this representation f depends on the thirteen independent quantities ρ, $\mathbf{u}$, T, p_{ij}, Q_i, so that the heat flux and stress tensor appear as independent variables. As we shall see, this will allow us to describe far more general phenomena then the conventional Chapman–Enskog (or Navier–Stokes) theory where the heat flux and stress tensor are expressed as gradients of the temperature or velocity field, respectively, but not both. In the thirteen-moment approximation we then write f as

$$f = f_{\text{LM}}(1 + \tfrac{1}{2}a_\alpha{}^{(2)}H_\alpha{}^{(2)} + b_\alpha H_{\alpha\beta\beta}{}^{(3)}) \tag{7-35}$$

where b_α is to be evaluated. Multiplying the above expression by $H_{\alpha\beta\beta}{}^{(3)}$ and integrating we find that $b_\alpha = a_\alpha{}^{(3)}/10$, so that

$$f = f_{\text{LM}}\left(1 + \frac{p_{\alpha\beta}}{2pRT}v_{0\alpha}v_{0\beta} - \frac{Q_\alpha v_{0\alpha}}{2pRT}\left(1 - \frac{v_0{}^2}{5RT}\right)\right) \tag{7-36}$$

The set of coefficient equations in the thirteen-moment approximation includes

$$\begin{aligned}
&\frac{\partial \rho}{\partial t} + \frac{\partial}{\partial \mathbf{q}}\cdot \rho\mathbf{u} = 0\\
&\frac{\partial u_\alpha}{\partial t} + \mathbf{u}\cdot\frac{\partial u_\alpha}{\partial \mathbf{q}} + \frac{1}{\rho}\frac{\partial}{\partial q_\beta}P_{\alpha\beta} = 0\\
&\frac{\partial p}{\partial t} + \frac{\partial}{\partial \mathbf{q}}\cdot \mathbf{u}p + \frac{2}{3}\mathsf{P} : \frac{\partial \mathbf{u}}{\partial \mathbf{q}} + \frac{1}{3}\frac{\partial}{\partial \mathbf{q}}\cdot \mathbf{Q} = 0
\end{aligned} \tag{7-37}$$

which are just the usual thermo-fluid conservation equations, plus two additional equations which, in the present scheme, replace the Fourier and Newton–Stokes laws in determining $\mathbf{Q}$ and P_{ij}. These additional two equations are

$$\begin{aligned}
\frac{\partial}{\partial t}p_{ij} &+ \frac{\partial}{\partial \mathbf{q}}\cdot \mathbf{u}p_{ij} + \frac{1}{5}\left(\frac{\partial Q_i}{\partial q_j} + \frac{\partial Q_j}{\partial q_i} - \frac{2}{3}\delta_{ij}\frac{\partial}{\partial \mathbf{q}}\cdot \mathbf{Q}\right)\\
&+ p_{i\alpha}\frac{\partial u_j}{\partial q_\alpha} + p_{j\alpha}\frac{\partial u_i}{\partial q_\alpha} - \frac{2}{3}\delta_{ij}p_{\alpha\beta}\frac{\partial u_\beta}{\partial q_\alpha}\\
&+ p\left(\frac{\partial u_i}{\partial q_j} + \frac{\partial u_j}{\partial q_i} - \frac{2}{3}\delta_{ij}\frac{\partial}{\partial \mathbf{q}}\cdot \mathbf{u}\right) + \beta p p_{ij} = 0
\end{aligned} \tag{7-38}$$

$$\begin{aligned}
\frac{\partial \mathbf{Q}}{\partial t} &+ \left(\frac{\partial}{\partial \mathbf{q}}\cdot \mathbf{u}\mathbf{Q}\right) + \left(\tfrac{7}{5}\mathbf{Q}\cdot\frac{\partial}{\partial \mathbf{q}}\right)\mathbf{u} + \tfrac{2}{5}Q_\alpha\frac{\partial}{\partial \mathbf{q}}u_\alpha\\
&+ \tfrac{2}{5}\mathbf{Q}\frac{\partial}{\partial \mathbf{q}}\cdot \mathbf{u} + 2RT\frac{\partial}{\partial \mathbf{q}}\cdot \mathbf{p} + 7\mathbf{p}\cdot\frac{\partial RT}{\partial \mathbf{q}}\\
&- 2\frac{\mathbf{p}}{\rho}\cdot\frac{\partial}{\partial \mathbf{q}}\cdot \mathsf{P} + 5p\frac{\partial RT}{\partial \mathbf{q}} + \tfrac{2}{3}\beta p\mathbf{Q} = 0
\end{aligned} \tag{7-39}$$

These last two equations are obtained from Equation (7–25) by noting that in the thirteen-moment approximation we can set

$$a_{ijk}^{(3)} = \tfrac{1}{5}(a_i^{(3)}\delta_{jk} + a_j^{(3)}\delta_{ik} + a_k^{(3)}\delta_{ij})$$
$$Q_{ijk} = \tfrac{1}{5}(Q_i\delta_{jk} + Q_j\delta_{ik} + Q_k\delta_{ij}) \tag{7–40}$$

The β which appears in Equations (7–38) and (7–39) is the contribution of the collision term, and, in general, this term will be a function of temperature. For Maxwell molecules, however, this term is a constant, since in this case $B(\theta,V) = B(\theta)$. In the concluding section we will show how β may be evaluated for Maxwell molecules.

In the next section we consider some of the implications of considering the heat flow and stress tensor as additional macroscopic variables.

7–5 SOME APPLICATIONS OF THE THIRTEEN-MOMENT EQUATIONS

Let us first consider two steady flows. In the case of a one-dimensional steady heat flow in a gas at rest we have, from the conservation equations,

$$\frac{\partial P_{ix}}{\partial q_x} = 0 \rightarrow P_{ix} = \text{constant}$$
$$\frac{\partial Q_x}{\partial q_x} = 0 \rightarrow Q_x = \text{constant} \tag{7–41}$$

From the additional two equations which replace the Fourier and Newton–Stokes laws we find

$$p_{ix} = 0 \rightarrow p = \text{constant}$$
$$\tfrac{5}{2}p\,\frac{\partial RT}{\partial q_x} + \tfrac{2}{3}\beta\rho Q_x = 0 \tag{7–42}$$

or

$$Q_x = -\kappa\,\frac{\partial T}{\partial q_x} \tag{7–43}$$

This last equation is just the Fourier heat law. The value of the thermal conductivity κ is the same as that found by using the Chapman–Enskog approximation [the particular approximation we refer to here is that given by Equation (6–79)]. This identification follows from the value found for β in the next section.

Now let us consider a one-dimensional steady Couette flow in the absence of a heat flow, that is, in addition to all the time derivatives and

$\mathbf{Q}$, the following terms are also set equal to zero: u_y, u_z, $(\partial u_x/\partial q_x)$, $(\partial u_z/\partial q_z)$. From the conservation equations we now obtain

$$\frac{\partial P_{ix}}{\partial q_i} = 0$$

$$\tfrac{2}{3}p_{xy}\frac{\partial u_x}{\partial q_y} = 0 \tag{7-44}$$

and the additional equation for p_{ij} gives

$$(p + p_{yy})\frac{\partial u_x}{\partial q_y} + \beta\rho p_{xy} = 0 \tag{7-45}$$

If we take $p_{yy} \ll p$, then we find

$$p_{xy} = -\mu\frac{\partial u_x}{\partial q_y} \tag{7-46}$$

and we recover the Newton–Stokes law with viscosity coefficient μ again equal to the Chapman–Enskog approximation. Thus, for conventional steady flow problems we see that the Navier–Stokes equations will give the proper thermo-fluid description.

As a simple example which illustrates both the application of the general flow equations in the thirteen-moment approximation and a physical significance of the constant β which appears in these equations, we now consider the case of a homogeneous unsteady flow. For this case the equations for the heat flow and stress tensor reduce to

$$\frac{\partial p_{ij}}{\partial t} + \beta\rho p_{ij} = 0$$

$$\frac{\partial Q_i}{\partial t} + \frac{2}{3}\beta\rho Q_i = 0 \tag{7-47}$$

The solution of these equations is

$$p_{ij} = p_{ij}(0)e^{-\beta\rho t}$$

$$Q_i = Q_i(0)e^{-2/3\beta\rho t} \tag{7-48}$$

and we see that $1/\beta\rho$ plays the role of a relaxation time which describes the decay of the initial heat flux and stress in the system. At long times (compared to $1/\beta\rho$) the Navier–Stokes equations are again valid. For short times, however, these equations are not valid, and the generalized flow equations (7–38) and (7–39) must be used instead. This result was originally obtained by Maxwell using the moment method described in the preceding section. It can be shown that, in general, when the molecular and macroscopic times are not separated (as is the case in any developing flow) then the Navier–Stokes equations cannot be expected to be valid, and a description in terms of a more general set of flow equations, for example, the thirteen-moment equations, must be used.

7–6 EVALUATION OF THE COLLISION TERMS

In this concluding section we show how the collision term β which appears in the thirteen-moment equations can be evaluated for Maxwell molecules. Some results for general interparticle potentials will also be given.

Two distinct collision terms appear in the thirteen-moment approximation, $J^{(2)}$ and $J^{(3)}$ [see Equation (7–25) for $n = 2, 3$]. In calculating these two quantities, which are themselves defined by Equations (7–27) to (7–32), we will need explicit expressions for the bracket terms $[H^{(2)}]$ and $[H^{(3)}]$. Making use of Equations (3–15) and (3–16) and Equation (7–4) we find, after a straightforward calculation,

$$[H^{(2)}] = 2\hat{V}^2 \cos^2 \theta \boldsymbol{\alpha}^2 - \hat{V} \cos \theta \boldsymbol{\alpha} \hat{\mathbf{V}}$$
$$[H^{(3)}] = \tfrac{1}{2}(\mathbf{v}_1 + \mathbf{v}_2)[H^{(2)}] \qquad (7\text{–}49)$$

Since $B(\theta,V)$ is independent of ϵ for all symmetrical potentials, we can do the necessary ϵ integration (all power law potentials, or any potential characteristic of structureless particles, for that matter, are symmetric). Thus [see Equation (7–29)] we find

$$I_\nu^{(n)} = \int d\theta\, B(\theta, \hat{V}(RT)^{1/2}) \int d\epsilon\, [H_\nu^{(n)}] \qquad (7\text{–}50)$$

and from Equation (7–49) we see that for $n = 2, 3$ we will have to consider integrals of the form

$$\hat{V}^i \int_0^{2\pi} d\epsilon\, \alpha^i = I_i \qquad i = 1, 2 \qquad (7\text{–}51)$$

Following Grad, we introduce a Cartesian coordinate system with coordinates $(V,0,0)$ for $\mathbf{V}$. Then (see Figure 2–3) $\boldsymbol{\alpha}$ is given by $(\cos \theta, \sin \theta \cos \epsilon, \sin \theta \sin \epsilon)$, and we have

$$\mathbf{I}_1 = 2\pi \cos \theta \hat{\mathbf{V}} \qquad (7\text{–}52)$$

$$\mathbf{I}_2 = \pi \hat{V}^2 \sin^2 \theta \boldsymbol{\delta} + \pi(2 \cos^2 \theta - \sin^2 \theta) \hat{V}^2 \qquad (7\text{–}53)$$

Accordingly we will have, using Equation (7–49),

$$\int_0^{2\pi} d\epsilon\, [H^{(2)}] = 2\pi \sin^2 \theta \cos^2 \theta (\hat{V}^2 \boldsymbol{\delta} - 3\hat{\mathbf{V}}^2) \qquad (7\text{–}54)$$

so that if

$$B(\hat{V}) \equiv B = \pi \int d\theta \sin^2\theta \cos^2\theta \, B(\theta, \hat{V}(RT)^{1/2}) \tag{7-55}$$

then the $I_\nu^{(n)}$ for $n = 2, 3$ are given by

$$\begin{aligned} I^{(2)} &= 2B(\hat{V}^2\boldsymbol{\delta} - 3\hat{\mathbf{V}}^2) \\ I^{(3)} &= \tfrac{1}{2}(\hat{\mathbf{v}}_1 + \hat{\mathbf{v}}_2)I^{(2)} \end{aligned} \tag{7-56}$$

The above results are for general symmetric interparticle potentials. For Maxwell molecules B is independent of $\hat{V}$, and we can directly evaluate $J^{(2)}$ and $J^{(3)}$. On the basis of our above results we find

$$\begin{aligned} J_{ij}^{(2)} &= -\beta\rho a_{ij}^{(2)} \\ J_i^{(3)} &= -\tfrac{2}{3}\beta a_i^{(3)} \end{aligned} \tag{7-57}$$

where $\beta = (6B/m)$. Making use of some standard trigonometric identities, we can then obtain the following relationship between B and the integral $A_2(5)$ which appears in the Chapman–Enskog theory:

$$B = \frac{\pi}{4}\left(\frac{2}{Km}\right)^{1/2} A_2(5) \tag{7-58}$$

The comparisons we made in the preceding sections between the results of the Chapman–Enskog theory and Grad's theory follow from the above equality.

For non-Maxwell molecules the integrations that must be done to evaluate $B(\hat{V})$ are complicated by the velocity dependence of this quantity. The closure conditions we have imposed limit the number of terms to be considered in the sum

$$J_\nu^{(n)} = \frac{\rho}{2m} \sum_{r,s=0} \beta_{\nu\rho\sigma}^{(nrs)} a_\rho^{(r)} a_\sigma^{(s)} \tag{7-59}$$

to $r, s \leq n$. The β's can be evaluated; however, this involves a great deal of laborious computation, and we shall not do this here. The results are of the form

$$\begin{aligned} J^{(2)} &= J_{MM}^{(2)} + O(a^2) \\ J^{(3)} &= J_{MM}^{(3)} + O(a^2) \end{aligned} \tag{7-60}$$

where $J_{MM}^{(i)}$ is the same functional of β and $a^{(i)}$ obtained for Maxwell molecules. Of course here the value of β will be different from that found for Maxwell molecules. Specifically, for power law potentials

$$\beta = \frac{6}{m\sqrt{\pi}\,5!} \int_0^\infty d\hat{V}\, \hat{V}^6 e^{-\hat{V}^2/4} \int_0^{2\pi} d\theta \sin^2\theta \cos^2\theta \, B\Big(\theta, \hat{V}(RT)^{1/2}\Big)$$

In addition there are correction terms which are quadratic in the a's, but for small gradients these terms may be neglected.

References

The Hermite tensor polynomials were introduced by:

1. H. Grad, *Commun. Pure and Appl. Math.*, vol. 2, p. 225, 1949.

Moment method expansions of Boltzmann's equation were proposed by:

2. H. Grad, *Commun. Pure and Appl. Math.*, vol. 2, p. 231, 1949.
3. J. C. Maxwell (1965). See General References.
4. D. Mintzer, in *Mathematics of Physics and Chemistry*, vol. 2, H. Margenau and G. Murphy, Eds. Princeton, N.J.: Van Nostrand, 1964.

A critical review of various moment methods is given by:

5. E. Ikenberry and C. Truesdell, *J. Rational Mech. and Analysis*, vol. 5, p. 1, 1956.

Problems

7–1. Write explicit expressions for $H_{ijkl}^{(4)}$ and $a_{ijkl}^{(4)}$.

7–2. Prove that

$$\frac{\partial}{\partial x_i} H^{(n)}(\mathbf{x}) = \delta_i H^{(n-1)}(\mathbf{x})$$

7–3. In Section 7–5 we showed that a simple one-dimensional heat flow did not produce stresses in a fluid. Show that this is not necessarily the case when the heat flow is two-dimensional, and explain why.

7–4. Show that the Newton–Stokes relationship is embedded in the thirteen-moment equations by considering the limit κ, $\mu \to 0$ ($\beta \to \infty$) in such a manner that all moments and derivatives are bounded.

7–5. Find the characteristic equation of the thirteen-moment equations which describe a one-dimensional flow, and discuss its properties.

7–6. Verify Equations (7–37) and (7–38).

7–7. Show that the Boltzmann collision term for Maxwell molecules can be written as

$$J(f) = \frac{1}{2m} \rho f_{\mathrm{LM}} \sum_{n=1} \sum_{r+s=n} \frac{1}{n!} \beta_{\nu\rho\sigma}{}^{(nrs)} a_\rho{}^{(r)} a_\sigma{}^{(s)} H_\nu{}^{(n)}$$

7–8. Show that the linearized collision operator for Maxwell molecules can be written as

$$L(g) = \frac{1}{m} \rho f_{\mathrm{LM}} \sum_{n=1} \beta_{\nu\sigma}{}^{(n0n)} a_\sigma{}^{(n)} H_\nu{}^{(n)}$$

CHAPTER 8

model boltzmann equations

8–1 THE SIMPLE BGK MODEL EQUATION

In attempting to utilize the Boltzmann equation to treat a particular flow problem we can expect to encounter many difficulties. In fact, there is only one example where Boltzmann's equation has been used to obtain an exact solution to a flow problem (and even here the problem is somewhat artificial). The difficulties one is faced with reduce to manageable proportions when the system to be considered is not too far removed from equilibrium, so that the linearized version of Boltzmann's equation can be used. However, even when the linearized Boltzmann equation is used, a great deal of painstaking labor is still involved in obtaining solutions for problems of interest. For these reasons Bhatnager, Gross, and Krook, and, independently, Welander, introduced a simplified model of the Boltzmann equation which can be used to treat general flow problems. In practice, the model leads to qualitatively correct results for a wide variety of flow situations; however, its simple structure also renders it inadequate in many respects. In order to improve the basic model, more complicated versions have been subsequently introduced (see Section 8–2). In turn, these models offer many of the problems found in the linearized Boltzmann equation itself, and it is debatable, despite their popularity, whether they do in fact offer a significant simplification of the treatment of flow problems.

The simple single relaxation time model used in kinetic theory is usually referred to as the BGK model, after Bhatnager, Gross, and Krook, who introduced the model in a paper published in 1954 (an identical model was independently proposed by Welander in a paper published the same year). The modeling is done by replacing the collision term $J(f)$ in Boltzmann's equation by the expression $\nu(f_{\mathrm{LM}} - f)$, so that in place of Boltzmann's equation we consider instead the model equation

$$\frac{\partial f}{\partial t} + \mathbf{v} \cdot \frac{\partial f}{\partial \mathbf{q}} = \nu(f_{\mathrm{LM}} - f) \tag{8–1}$$

The quantity ν is the collision frequency (reciprocal relaxation time), and it is usually treated as an adjustable parameter, independent of the variables $\mathbf{q}$, $\mathbf{v}$, t which appear in the argument of f.

Let us consider how we may view the BGK model equation as being "derived" from Boltzmann's equation. If we assume the system being treated is not too far from equilibrium, we may then replace the f's in the positive part of the collision term by their local equilibrium values, so that this part of the collision term can be written as

$$\frac{1}{m}\int d\mathbf{v}_2\, d\theta\, d\epsilon\, B(\theta,V) f_1' f_2' \rightarrow \frac{1}{m}\int d\mathbf{v}_2\, d\theta\, d\epsilon\, B(\theta,V) f_{1_{\mathrm{LM}}}' f_{2_{\mathrm{LM}}}'$$

$$= \frac{1}{m}\int d\mathbf{v}_2\, d\theta\, d\epsilon\, B(\theta,V) f_{1_{\mathrm{LM}}} f_{2_{\mathrm{LM}}} \equiv \nu_{\mathrm{L}}(\mathbf{v}_1) f_{1_{\mathrm{LM}}} \tag{8-2}$$

where [see Equation (4–35)]

$$\nu_{\mathrm{L}}(\mathbf{v}_1) = \frac{1}{m}\int d\mathbf{v}_2\, d\theta\, d\epsilon\, B(\theta,V) f_{2_{\mathrm{LM}}} \tag{8-3}$$

Usually, however, ν_{L} is not considered as a function of velocity. This is equivalent to treating the particles as Maxwell molecules, that is, taking $B(\theta,V)$ to be a function of θ only. In this case

$$\int d\mathbf{v}_2\, d\theta\, d\epsilon\, B(\theta)\{f_2 - f_{2_{\mathrm{LM}}}\} = 0 \tag{8-4}$$

and the entire collision term may be replaced by the term on the right-hand side of Equation (8–1):

$$J(f) \rightarrow \nu(f_{\mathrm{LM}} - f) \tag{8-5}$$

This is the simplest way of "deriving" the BGK equation. The same equation may also be extracted from the linearized Boltzmann equation in the following way. Using the results of the last chapter (specifically Section 7–4) we may write the full collision operator as

$$J(f) = \tfrac{1}{2}\rho f_{\mathrm{LM}} \sum_{n=1}^{\infty} \sum_{r+s=n} \frac{1}{n!} \beta_{\nu\rho\sigma}{}^{(nrs)} a_\rho{}^{(r)} a_\sigma{}^{(s)} H_\nu{}^{(n)} \tag{8-6}$$

[This form of $J(f)$ is implicit in Equation (7–25).] In considering the linearized collision operator for Maxwell molecules we retain only the two terms $r = 0$, $s = n$ and $r = n$, $s = 0$ in the above sum so that the linearized version of Equation (8–6) is given by

$$L(f) = \rho f_{\mathrm{LM}} \sum_{n=1}^{\infty} \beta_{\nu\sigma}{}^{(n0n)} a_\sigma{}^{(n)} H_\nu{}^{(n)} \tag{8-7}$$

If we now write $-\rho\beta_{ik}^{(n0n)} = \nu\delta_{ik}$ and substitute back into Equation (8–7) we recover the BGK collision term. This assumption is certainly *ad hoc;* it amounts to retaining only one collision time in the system.

The simple BGK equation can be treated in much the same way as the Boltzmann equation. Normal solutions may be found, and multi-moment expansions carried out. It should be pointed out that the equation is not as simple as it may appear at first sight. It is strongly non-linear since f_{LM} contains ρ, $\mathbf{u}$, T, which are defined in terms of moments of f. Among the properties that the simple BGK equation has in common with the Boltzmann equation are an H theorem, and the correct prediction of the thermo-fluid conservation equations. One of its most glaring shortcomings is that it gives the wrong value for the ratio μ/κ (this is usually called the Prandtl number).

As a trivial example of the application of the BGK equation let us consider the initial value problem for a space independent flow. We have already seen that this is a difficult problem to treat with the linearized Boltzmann equation; however, the exact solution to Equation (8–1) is easily found, namely,

$$f(\mathbf{v},t) = f_0(\mathbf{v})e^{-\nu t} + (1 - e^{-\nu t})f_{LM}(\mathbf{v}) \tag{8–8}$$

The ρ, $\mathbf{u}$, T which appear in f_{LM} can be directly calculated in terms of the first five moments of the initial distribution function, f_0. We see that any given initial distribution function decays to a Maxwellian, with a characteristic relaxation time ν^{-1}.

In practice it is often desirable to treat a simplified version of the BGK equation which is obtained from Equation (8–1) by linearization. We proceed by writing

$$f = f_0(1 + \phi) \tag{8–9}$$

so that

$$\begin{aligned} \rho &= \rho_0(1 + \bar{\rho}) \\ \mathbf{u} &= \bar{\mathbf{u}} \\ T &= T_0(1 + \bar{T}) \end{aligned} \tag{8–10}$$

with

$$\begin{Bmatrix} \rho_0 \\ 3\rho_0 RT_0 \end{Bmatrix} = \int d\mathbf{v}\, f_0 \begin{Bmatrix} 1 \\ v^2 \end{Bmatrix} \tag{8–11}$$

and

$$\begin{Bmatrix} \bar{\rho}\rho_0 \\ \rho_0\bar{\mathbf{u}} \\ 3\rho_0 R\bar{T}T_0 \end{Bmatrix} = \int d\mathbf{v}\, f_0\phi \begin{Bmatrix} 1 \\ \mathbf{v} \\ v^2 - 3RT_0 \end{Bmatrix} \tag{8–12}$$

The linearized form of f_{LM} is obtained by substituting Equation (8–10) into the local Maxwellian distribution function and retaining only those terms which are not quadratic, or of higher degree, in the perturbations $\bar{\rho}$, $\bar{\mathbf{u}}$, $\bar{T}$. Thus we will have

$$f_{LM} = f_0\left\{1 + \bar{\rho} + \frac{\mathbf{v}_0 \cdot \bar{\mathbf{u}}}{RT_0} + \bar{T}\left(\frac{v^2}{2RT_0} - \frac{3}{2}\right)\right\} \tag{8-13}$$

where

$$f_0 = \frac{\rho_0}{(2\pi RT_0)^{3/2}} e^{-\mathbf{v}^2/2RT_0} \tag{8-14}$$

Accordingly, in this approximation, the BGK equation is replaced by

$$\frac{\partial \phi}{\partial t} + \mathbf{v} \cdot \frac{\partial \phi}{\partial \mathbf{q}} = \nu\left\{-\phi + \bar{\rho} + \frac{\mathbf{v}_0 \cdot \bar{\mathbf{u}}}{RT_0} + \bar{T}\left(\frac{v^2}{2RT_0} - \frac{3}{2}\right)\right\} \tag{8-15}$$

8-2 GENERALIZED MODEL EQUATIONS

In order to obtain a model equation which imitates Boltzmann's equation more closely than the simple BGK model does, it is necessary to incorporate more structure into its construction. For example, if additional adjustable parameters are included, then the correct Prandtl number can be obtained. In a certain sense this can be considered as a self-defeating procedure, since the primary virtue of a model equation lies in its inherent simplicity of structure. In the present section we consider a scheme, originally proposed by Gross and Jackson for Maxwell molecules, and subsequently generalized by Sirovich for arbitrary interparticle potentials, which prescribes an infinite set of model equations of increasing complexity.

Consider the linearized Boltzmann collision operator, $L(g)$, for Maxwell molecules. The eigenfunctions of this operator constitute a complete set, so that we may expand g in this set, and obtain the following expression for $L(g)$:

$$L(g) = \sum_i a_i \lambda_i \psi_i \tag{8-16}$$

Here the λ_i are the eigenvalues of L, corresponding to the eigenfunctions ψ_i, and the a_i are constants which are determined using the orthogonality properties of the ψ_i. In the Gross–Jackson scheme the spectrum of L is mutilated by representing all eigenvalues λ_i for $i > N$ by the single value $-\nu_N$. Equivalently, we can replace the operator L by the new operator L_N, which is defined as

$$L_N(g) = \sum_{i=1}^{N} a_i \lambda_i \psi_i - \nu_N \sum_{i=N+1}^{\infty} a_i \psi_i \tag{8-17}$$

The above equation can be rewritten to read

$$L_N(g) = \sum_{i=1}^{N} (\nu_N a_i + a_i \lambda_i)\psi_i - \nu_N g \tag{8-18}$$

Since $\lambda_i = 0$ for $i = 1 - 5$, we see that the choice $N = 5$ corresponds to retaining only one relaxation time in the system. The model equation which results is just the linearized version of the simple BGK equation when ν_N is set equal to ν. To see this we note that the first five eigenfunctions of L are

$$1, \frac{\mathbf{v}}{\sqrt{RT}}, \sqrt{3/2}\left(1 - \frac{v^2}{3RT}\right)$$

so that taking $N = 5$ in Equation (8–18), we find

$$\frac{\partial g}{\partial t} + \mathbf{v} \cdot \frac{\partial \mathbf{q}}{\partial g} = \nu \left\{ -g + \bar{\rho} + \frac{\mathbf{v} \cdot \bar{\mathbf{u}}}{RT_0} + \bar{T}\left(\frac{v^2}{2RT_0} - \frac{3}{2}\right)\right\} \tag{8-19}$$

since [see Equation (4–39)]

$$\begin{aligned} a_1 &= \bar{\rho} \\ a_{2,3,4} &= \bar{u}_{x,y,z}/\sqrt{RT} \\ a_5 &= -\sqrt{\tfrac{3}{2}}\,\bar{T} \end{aligned} \tag{8-20}$$

The next Gross–Jackson equation which is of particular interest is that found by choosing $N = 7$. In addition to ν_7, this equation also contains λ_6 and λ_7 as adjustable parameters. This additional flexibility allows us to correctly describe the Navier–Stokes equations and the thirteen-moment equations. We shall consider this model in the Problems at the end of this chapter.

8–3 SOME OTHER MODEL EQUATIONS

In concluding this chapter on model equations we would like to mention some models which have been introduced to treat specific technical problems in kinetic theory. The first, due to Kac, is a simple caricature of Boltzmann's equation which has been used by both Kac and McKean to demonstrate some not so simple technical properties of the equation. The model is

$$\frac{\partial f}{\partial t}(v_1,t) = \frac{1}{2\pi}\int_{-\infty}^{\infty} dv_2 \int_0^{2\pi} d\theta \\ [f(v_1 \cos\theta - v_2 \sin\theta, t) f(v_1 \sin\theta + v_2 \cos\theta, t) - f(v_1,t)f(v_2,t)] \tag{8-21}$$

We see that Kac's equation describes a one-dimensional, spatially homogeneous system in which binary collisions take the state $(v_1 \cos\theta - v_2 \sin\theta,\ v_1 \sin\theta + v_2 \cos\theta)$ into (v_1,v_2). Note that "collisions" in this model conserve only number and energy, and not momentum.

Another model, perhaps even simpler than that of Kac, which has also been used to prove some complicated technical properties, is the two-state model proposed by Carleman. This is a purely mathematical model, and consists of the two equations

$$\begin{aligned} \frac{\partial n_+}{\partial t} + v\frac{\partial n_+}{\partial x} &= n_-^2 - n_+^2 \\ \frac{\partial n_-}{\partial t} - v\frac{\partial n_-}{\partial x} &= n_+^2 - n_-^2 \end{aligned} \tag{8–22}$$

Here n_+ and n_- can be considered as the distribution functions for particles moving with speed v in the positive and negative directions along a line. A generalization of this model by Broadwell to three dimensions provides yet another model. This model has been successfully employed in treating complicated boundary value problems. In constructing the generalization of Carleman's model, it is assumed that the particles can move with only certain velocities, $\mathbf{v}_i$; if n_i particles per unit volume have velocity $\mathbf{v}_i$, then the model Boltzmann equation is

$$\frac{\partial n_i}{\partial t} + \mathbf{v}_i \cdot \frac{\partial n_i}{\partial \mathbf{q}_i} = \sum_{j,k,l} |\mathbf{v}_k - \mathbf{v}_l| A_{kl}^{ij} n_k n_l - \sum_{j,k,l} |\mathbf{v}_i - \mathbf{v}_j| A_{ij}^{kl} n_i n_j \tag{8–23}$$

Here A_{kl}^{ij} is the effective collision cross section for a binary collision taking the state (v_k,v_l) into the state (v_i,v_j). For spherically symmetric interparticle potentials $A_{kl}^{ij} = A_{ij}^{kl}$. As an example of this discrete velocity gas we consider a system in which the particles are constrained to move along the axis of a standard three-dimensional Cartesian coordinate system at constant speed v, so that only six velocities are permissible. If the forces which act between the particles are spherically symmetric, and if we require collisions between the particles to conserve particles and momentum (particle conservation implies energy conservation in this model), then the latter requirement implies that only collisions between pairs of particles having (v_x,v_{-x}), (v_y,v_{-y}), and (v_z,v_{-z}) are effective in changing the n_i. Since the interparticle potential is spherically symmetric we must weigh the three possible outcomes of each of these head-on collisions equally, that is,

$$(v_x,v_{-x}) \xrightarrow[\text{collision}]{\text{binary}} \begin{matrix} (v_x,v_{-x}) \\ (v_y,v_{-y}) \\ (v_z,v_{-z}) \end{matrix} \tag{8–24}$$

each with probability $\frac{1}{3}$, so that, for example, if σ is the mutual binary collision cross section for these head-on collisions, then $\sigma/3$ is the effective collision cross section, and the Boltzmann equation for this model is

$$\begin{aligned}
\frac{\partial n_x}{\partial t} + v\frac{\partial n_x}{\partial x} &= \frac{2}{3}\sigma v[n_y n_{-y} + n_z n_{-z} - 2n_x n_{-x}] \\
\frac{\partial n_{-x}}{\partial t} - v\frac{\partial n_{-x}}{\partial x} &= \frac{2}{3}\sigma v[n_y n_{-y} + n_z n_{-z} - 2n_x n_{-x}] \\
\frac{\partial n_y}{\partial t} + v\frac{\partial n_y}{\partial y} &= \frac{2}{3}\sigma v[n_x n_{-x} + n_z n_{-z} - 2n_y n_{-y}]
\end{aligned} \tag{8-25}$$

and so forth.

In the following chapter we shall see how the models introduced in this chapter can be utilized to solve some particular flow problems of interest.

References

The simple BGK model equation was first used by:

1. P. L. Bhatnager, E. P. Gross, and M. Krook, *Phys. Rev.*, vol. 94, p. 511, 1954.
2. P. Welander, *Arkiv. Fysik.*, vol. 7, p. 507, 1954.

The generalized BGK equations are due to:

3. C. Cercignani, *Ann. Phys.*, vol. 40, p. 469, 1966.
4. E. P. Gross and E. A. Jackson, *Phys. Fluids*, vol. 2, p. 432, 1959.
5. L. Sirovich, *Phys. Fluids*, vol. 5, p. 908, 1962.

Some other model Boltzmann equations have been considered by:

6. T. Carleman (1957). See General References.
7. J. Broadwell, *J. Fluid Mech.*, vol. 19, p. 401, 1964.
8. M. Kac, *Probability and Related Topics in the Physical Sciences.* New York: Interscience, 1959.
9. A. Povzner, *Mat. Sb.*, vol. 58, p. 62, 1962.

Problems

8–1. Show that the simple BGK equation for a space-uniform system satisfies an H theorem.

8–2. Derive the usual equations of mass, momentum, and energy conservation from the simple BGK model equation.

8–3. Determine the Chapman–Enskog solution to the simple BGK equation in first and second approximations.

8–4. Using the results of Problem 8–3, determine κ and μ as calculated from the simple BGK equation in the second Chapman–Enskog approximation, and compare the ratio μ/κ with the value obtained from Boltzmann's equation in the corresponding approximation.

8–5. Suppose we wish to derive BGK-like model equations for a system of non-Maxwell molecules. Show that by expanding g in the eigenfunctions of L for Maxwell molecules we can, by analogy with the Gross–Jackson procedure, obtain the following model equation:

$$\frac{\partial g}{\partial t} + \mathbf{v} \cdot \frac{\partial g}{\partial \mathbf{q}} = -\nu g + \sum_{i,j=1}^{N} (\lambda_{ij} + \delta_{ij}\nu) a_i \psi_j$$

where the a_i are the expansion coefficients, and λ_{ij} is to be determined.

8–6. Despite the fact that Kac's model does not conserve momentum in collisions, we still find that $d\mathbf{v}_1\, d\mathbf{v}_2 = d\mathbf{v}_1'\, d\mathbf{v}_2'$. Prove this, and use this result to prove an H theorem for Kac's model Boltzmann equation (8–21).

8–7. Prove that the simple model equation (8–25) has an H theorem.

8–8. Derive the conservation equations for the simple model equation (8–25).

CHAPTER 9

solutions of special problems

9-1 SOLUTIONS OF BOLTZMANN'S EQUATION

The solutions of the Boltzmann (and model) equation(s) which may be found in the literature fall into two basic categories. The first type of solution is of a formal nature. These are mathematically rigorous proofs of the existence of solutions to Boltzmann's equation for particular problems. The second type is of a less general, and more *ad hoc*, nature. These are specific solutions for particular initial and boundary value problems of interest, and they are almost never exact solutions. As might be expected, preoccupation with the first type of result has been confined mostly to mathematicians, while the second type of result has been of interest mainly to engineers and physicists. Since the latter are not, as a rule, adverse to providing a specific result without first demonstrating rigorous existence proofs, it is not surprising that in many cases a result of the second type has been obtained for a problem where a rigorous existence proof is lacking.

In the present chapter we will be concerned with results of the second type. Most of these results have been obtained in the past fifteen years, a period which has seen a great revival of interest in the kinetic theory. A product of this increasing interest is a literature which appears to be growing at an ever accelerating rate. Much of this literature appears, either as a contribution or in reference, in the *Proceedings of the Rarefied Gas Dynamics Symposium*, which has been issued every two years since the inception of the Symposium in 1958. Both the increasing interest in kinetic theory and the international composition of the workers in this field are evident from the growing contents of these proceedings.

In the sections that follow we will consider some of the basic approaches used in obtaining solutions to Boltzmann's equation and its related model equations for some specific flow problems of interest. We will be most interested in the various methods used in attacking specific types of problems rather than in all the details of the actual solution, which may be found in the literature cited. As examples we will consider Poiseuille flow, light scattering, shock wave structure, and heat and shear flow. These problems have been singled out by us as representative of the types of problems which have been treated in the literature. Before considering these specific problems, however, it seems worthwhile to at least mention the important formal existence results which have also been obtained.

The first results of this nature appear to be due to Carleman. Carleman was able to show the existence of solutions to Boltzmann's equation for the case of hard sphere molecules when the distribution function is spatially uniform and isotropic in the velocity space (that is, a function of v rather than $\mathbf{v}$). This result was later generalized by both Morgenstern and Wild for distribution functions having full velocity dependence; however, their results are only valid for cutoff Maxwell molecules. For spatially dependent systems Carleman obtained weak existence proofs, valid only for very short times, for both the full and linearized Boltzmann equation; these results are also only for hard sphere molecules. Grad has generalized these results to cutoff power law potentials, and also improved them, extending them for long times (in the case of the full Boltzmann equation this improvement is only for a restricted class of initial conditions). Only one result exists for infinite range potentials, this being for the spatially uniform, linearized Boltzmann equation for Maxwell molecules.

9–2 BOUNDARY CONDITIONS IN KINETIC THEORY

In considering flow problems in kinetic theory we will quite often have to take into account the interaction of the system particles with a solid boundary, for example, as in the flow of a rarefied gas between two moving plates (Couette flow). In order to treat this problem we will introduce a function $K_w(\mathbf{v}',\mathbf{v};\mathbf{q})$ which gives the probability that a particle which strikes the wall at $\mathbf{q}$ with velocity $\mathbf{v}'$ will emerge at $\mathbf{q}$ with velocity $\mathbf{v}$. We will assume that the particles are not captured by the wall for any duration, but are instantaneously reemitted upon impinging. If we denote the unit normal drawn from the wall to the system as $\hat{\mathbf{n}}$, then for $\mathbf{v}' \cdot \hat{\mathbf{n}} < 0$ we have

$$\int_{\mathbf{v}\cdot\hat{\mathbf{n}}>0} d\mathbf{v}\, K_w(\mathbf{v}',\mathbf{v};\mathbf{q}) = 1 \tag{9–1}$$

It then follows from mass conservation at the wall that

$$\int d\mathbf{v}\, \mathbf{v} \cdot \hat{\mathbf{n}} f(\mathbf{q},\mathbf{v},t) = 0 \tag{9–2}$$

The problem of specifying the form of K_w was already considered by Maxwell, who distinguished between two basic types of particle–wall interactions. In the first, a particle strikes the wall and is specularly reflected, so that $\mathbf{v} = \mathbf{v}' - 2\hat{\mathbf{n}}(\hat{\mathbf{n}} \cdot \mathbf{v}')$ for all $\mathbf{v}' \cdot \hat{\mathbf{n}} < 0$, that is, only the component of $\mathbf{v}'$ normal to the wall is changed, and this is just reversed in direction. The second type of particle–wall interaction considered by Maxwell represents the opposite extreme, complete loss of memory by the reflected particle. In this case the reflected particle is expelled from the wall with a random velocity, with the probability of a given velocity of expulsion given by the local Maxwellian distribution function in which the temperature that appears is that of the wall. Of course this velocity must be directed back into the system. This case is usually referred to as diffuse reflection. A more general model of the particle–wall interaction is obtained by considering that some fraction, α, of the particles are specularly reflected, and $(1 - \alpha)$ are diffusely reflected. Then we will have

$$K_w(\mathbf{v}',\mathbf{v};\mathbf{q}) = \alpha\delta(\mathbf{v}' - \mathbf{v} + 2\hat{\mathbf{n}}(\hat{\mathbf{n}} \cdot \mathbf{v})) + (1 - \alpha)\frac{\rho_w(\mathbf{v} \cdot \hat{\mathbf{n}})}{(2\pi RT_w)^{3/2}} e^{-v^2/2RT_w} \quad \text{(9–3)}$$

By substituting this expression into Equation (9–1) we can evaluate $\rho_w = (2\pi/RT_w)^{1/2}$. In comparison with experiments it is found that although Equation (9–3) has definite shortcomings, this model with $\alpha \approx 0$ gives reasonable answers for most applications. As more results become available in the area of gas–surface interactions, it is clear that we will have to alter the Maxwell model, but until that time it will probably continue to be quite popular due to its simple structure and resulting ease of application.

9–3 PLANE POISEUILLE FLOW

As an introduction to the solution of flow problems in kinetic theory we first consider the relatively simple problem of a plane Poiseuille flow induced by a weak pressure gradient. Specifically, we consider a steady flow of a rarefied gas between two stationary, unheated parallel walls (see Figure 9–1). The wall temperature is T_w. The distance between the walls is d, and this separation must be large compared to the mean free path in order for the Boltzmann equation to be applicable. Accordingly, we consider that case in what follows.

In treating the Poiseuille flow problem described above, we will utilize the linearized BGK equation. The distribution function $f(\mathbf{q},\mathbf{v})$ is thus linearized according to

$$f(\mathbf{q},\mathbf{v}) = f(x,y,v_x,v_y) = \frac{\rho_0}{(2\pi RT_w)^{3/2}} e^{-\mathbf{v}^2/2RT_w}(1 + \phi) \equiv f_0(1 + \phi) \quad \text{(9–4)}$$

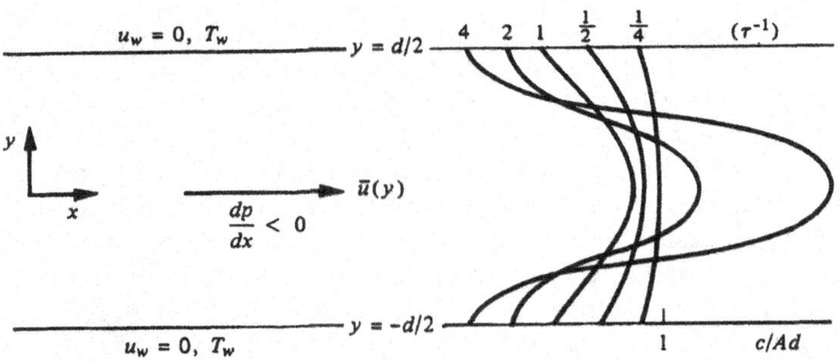

Figure 9–1 Flow geometry for Poiseuille flow and the velocity profiles as calculated from IX-18.

where the pertubation quantity ϕ satisfies the following equation:

$$\mathbf{v}\cdot\frac{\partial\phi}{\partial\mathbf{q}} = \nu\left[-\phi + \bar{p} + \frac{\mathbf{v}\cdot\bar{\mathbf{u}}}{RT_w} + \bar{T}\left(\frac{v^2}{2RT_w} - \frac{3}{2}\right)\right] \tag{9–5}$$

This equation is to be considered subject to boundary conditions, which we impose at each of the walls. Adopting the conditions of purely diffusive reflection, we must then require

$$f\left(x, \frac{d}{2}, \mathbf{v}\right) = \frac{\rho_w}{(2\pi RT_w)^{3/2}}\, e^{-v^2/2RT_w} \qquad v_y < 0$$
$$f\left(x, \frac{-d}{2}, \mathbf{v}\right) = \frac{\rho_w}{(2\pi RT_w)^{3/2}}\, e^{-v^2/2RT_w} \qquad v_y > 0 \tag{9–6}$$

It will prove convenient in what follows to introduce the following dimensionless variables into Equation (9–5):

$$\begin{aligned} \mathbf{w} &= \mathbf{v}(2RT_w)^{-1/2} \\ \mathbf{c} &= c\hat{\mathbf{i}} = \bar{\mathbf{u}}(2RT_w)^{-1/2} \\ \bar{\mathbf{q}} &= (X,Y) = \mathbf{q}d^{-1} = \left(\frac{x}{d}, \frac{y}{d}\right) \\ \tau^{-1} &= \nu d(2RT_w)^{-1/2} \end{aligned} \tag{9–7}$$

so that equation becomes

$$\tau\mathbf{w}\cdot\frac{\partial\phi}{\partial\bar{\mathbf{q}}} = -\phi + \bar{p} + 2\mathbf{w}\cdot\mathbf{c} + \bar{T}(w^2 - \tfrac{3}{2}) \tag{9–8}$$

A simple argument now allows us to deduce the form of ϕ and proceed directly to a solution. We note that, due to the assumed linear dependence of the small pressure gradient, we can write the pressure as

$$p = p_0(1 + \bar{p}) = \rho_0 RT_w(1 + \bar{p}) = \rho_0 RT_w(1 - Ax) \tag{9–9}$$

where A is a constant. But we can also write the pressure by expressing it as a moment of f in the usual manner,

$$p = \frac{1}{3}\int d\mathbf{v}\, f(\mathbf{v} - \bar{\mathbf{u}})^2 = p_0 + \frac{1}{3}\int d\mathbf{v}\, f_0\phi\mathbf{v}^2 \tag{9–10}$$

Since the two expressions for $\bar{p}$ must be equal, we see by comparing them that ϕ must be of the following form:

$$\phi = -Ax + w_x\psi(\bar{\mathbf{q}},\mathbf{w}) \tag{9–11}$$

An equation for ψ can be obtained by substituting the above expression into Equation (9–8). Doing this, and simplifying by substituting for $\bar{p}$ and $\bar{T}$, we find

$$\tau w_x \frac{\partial\psi}{\partial X} + \tau w_y \frac{\partial\psi}{\partial Y} - A\tau d = -\psi + 2c \tag{9–12}$$

The boundary conditions on ψ follow from those on f, plus Equations (9–4) and (9–11), which give

$$\begin{aligned} \psi(X,\tfrac{1}{2},\mathbf{w}) &= 0 \qquad w_y < 0 \\ \psi(X,-\tfrac{1}{2},\mathbf{w}) &= 0 \qquad w_y > 0 \end{aligned} \tag{9–13}$$

Inspecting Equations (9–12) and (9–13) we see that X nowhere appears explicitly, and therefore ψ does not contain that variable explicitly. Therefore we can simplify Equation (9–12) by dropping the $\partial/\partial X$ term, and then rewriting that differential equation as the following pair of integral equations:

$$\begin{aligned} \psi(Y,\mathbf{w}) &= \tau^{-1}\int_{-1/2}^{Y} \frac{d\xi}{w_y}\,[\tau A d + 2c(\xi)]e^{-\tau^{-1}(Y-\xi)/w_y} \qquad w_y > 0 \\ \psi(Y,\mathbf{w}) &= \tau^{-1}\int_{+1/2}^{Y} \frac{d\xi}{w_y}\,[\tau A d + 2c(\xi)]e^{-\tau^{-1}(Y-\xi)/w_y} \qquad w_y < 0 \end{aligned} \tag{9–14}$$

The flow field c is then

$$\begin{aligned} c &= \pi^{-3/2}\int d\mathbf{w}\, e^{-\mathbf{w}^2}\psi w_x^2 \\ &= \frac{\tau^{-1}}{\pi^{1/2}}\int_{-1/2}^{1/2} d\xi \left[\frac{\tau A d}{2} + c(\xi)\right] J_{-1}(\tau^{-1}|Y - \xi|) \end{aligned} \tag{9–15}$$

The J function which appears in the above expression is a particular example of the so-called Abramowitz integrals which appear quite often in the kinetic theory of gases, and have been tabulated. The general integral, $J_n(z)$, is defined as

$$J_n(z) = \int_0^\infty ds\, s^n e^{-(s^2+zs^{-1})} \tag{9–16}$$

The particular integral which appears in Equation (9–15) has a singularity at $\xi = Y$, so that we may effect a very good approximation for c by making use of the fact that the weight of the integrand will be concentrated about that singularity. Making use of the relationship

$$J_n(z) = -\frac{d}{dz} J_{n+1}(z) \tag{9–17}$$

we then find

$$c \approx \frac{\tau A d}{2}\left[\frac{\sqrt{\pi} - J_0(\tau^{-1}(\frac{1}{2} - Y)) - J_0(\tau^{-1}(\frac{1}{2} + Y))}{J_0(\tau^{-1}(\frac{1}{2} - Y)) + J_0(\tau^{-1}(\frac{1}{2} + Y))}\right] \tag{9–18}$$

This solution exhibits the well-known Knudsen paradox, first observed experimentally by Martin Knudsen in 1909. The paradox refers to the unexpected result that the flow rate

$$Q = d\int_{-1/2}^{1/2} dY\, c(Y)$$

is a minimum for some value of the pressure. This can be seen by noting that $c(Y) \to \infty$ for both $\tau^{-1} \to 0$ and $\tau^{-1} \to \infty$ and thus for some intermediate value of τ^{-1}, corresponding to a unique pressure, $c(Y)$, and hence the flow rate will go through a minimum.

9–4 THE DENSITY–DENSITY CORRELATION FUNCTION

In the preceding section we considered the solution of the simple, linearized model equation for a pure boundary value problem. The method of solution was fairly straightforward, and is illustrative of the approach which can often be used in effecting solutions to this class of flow problem. In the present section we consider the same model equation, but now we consider a problem which has both space and time dependence. We will again make use of a solution technique which has considerable application, in this case for time dependent flows. To illustrate this technique we will consider the calculation of the van Hove density–density correlation function, $G(\mathbf{q},t)$, which is of great interest in nonequilibrium statistical mechanics. The function $G(\mathbf{q},t)$ is used to measure the inelastic scattering of light or neutrons due to fluctuations of density and temperature which occur in a fluid. These results can be compared with experimental values, and thereby serve to provide us with a test of the model equation for frequencies on the order of a collision frequency and wavelengths on the order of a mean free path. Such experiments have recently begun to be carried out, and appear to be considerably more reliable than sound wave experiments, which were formerly used to provide a similar test.

The van Hove density–density correlation function, $G(\mathbf{q},t)$, can be interpreted as the probability per unit volume of finding a particle at $\mathbf{q}$ at time t given that there was a particle, not necessarily the same one, at the origin at $t = 0$. The quantity that is measured experimentally is not $G(\mathbf{q},t)$, but rather its Fourier–Laplace transform

$$S(\mathbf{K},\omega) = \int_{-\infty}^{\infty} dt \int d\mathbf{q}\, G(\mathbf{q},t)e^{i(\mathbf{K}\cdot\mathbf{q}-\omega t)} \tag{9–19}$$

In order to use our previous notation without change, we will assume in what follows that the mass is expressed in units in which each particle has a mass of unity.

From the above discussion we see that in a rarefied gas the problem of determining $G(\mathbf{q},t)$ is equivalent to finding $\rho(\mathbf{q},t)$ by solving the Boltzmann equation subject to the initial condition $\rho(0,0) = 1$. In practice we cannot solve Boltzmann's equation, so we use the linearized equation or, as in what follows, a model. The linearization is $f = f_M(1 + \phi)$, and the initial condition is then

$$\phi(\mathbf{q},\mathbf{v},0) = \phi(\mathbf{q},0) = \delta(\mathbf{q}) \tag{9–20}$$

For our present purpose it is not worthwhile to discuss at length why the initial pertubation is Maxwellian in the velocity space. In brief, this is a result of $G(\mathbf{q},t)$ being defined as an equilibrium ensemble average; for what follows it will suffice to consider Equation (9–20) as being given. We then have

$$G(\mathbf{q},t) = \rho_0(1 + \bar{\rho}(\mathbf{q},t)) \tag{9–21}$$

with $\bar{\rho}$ to be determined by solving the linearized BGK equation.

We proceed by taking the Fourier–Laplace transform of the linearized BGK equation (8–15) subject to Equation (9–20), which gives

$$(\nu + i\omega - i\mathbf{K}\cdot\mathbf{v})\tilde{\phi}(\mathbf{K},\mathbf{v},\omega) \equiv \Omega\tilde{\phi} = \nu\left[\tilde{\bar{\rho}} + \frac{\mathbf{v}\cdot\tilde{\bar{\mathbf{u}}}}{RT_0} + \tilde{\bar{T}}\left(\frac{v^2}{2RT_0} - \frac{3}{2}\right)\right] + S(\mathbf{K}) \tag{9–22}$$

where the tilde ($\tilde{\bar{\ }}$) denotes a Fourier–Laplace transform, and $S(\mathbf{K})$ is the Fourier–Laplace transformed initial condition. This equation can be solved for $\tilde{\phi}$, and we find

$$\tilde{\phi}(\mathbf{K},\mathbf{v},\omega) = \frac{\nu}{\Omega}\left[\tilde{\bar{\rho}} + \frac{\mathbf{v}\cdot\tilde{\bar{\mathbf{u}}}}{RT_0} + \tilde{\bar{T}}\left(\frac{v^2}{2RT_0} - \frac{3}{2}\right)\right] + S(\mathbf{K}) \tag{9–23}$$

A set of equations for $\bar{\bar{\rho}}$, $\bar{\bar{\mathbf{u}}}$, and $\bar{\bar{T}}$ can be obtained from Equation (9–23) by multiplying this equation by f_M, $f_M\mathbf{v}$, $f_M(v^2 - \frac{3}{2})$ and integrating over $\mathbf{v}$. For instance, carrying out the first operation gives

$$\rho_0\bar{\bar{\rho}} = \nu\bar{\bar{\rho}}\int d\mathbf{v}\, f_M\Omega^{-1} + \frac{\nu\bar{\bar{\mathbf{u}}}}{RT_0}\cdot\int d\mathbf{v}\,\mathbf{v}f_M\Omega^{-1}$$
$$+ \nu\bar{\bar{T}}\int d\mathbf{v}\, f_M\left(\frac{v^2}{2RT_0} - \frac{3}{2}\right)\Omega^{-1} + S(\mathbf{K})\int d\mathbf{v}\, f_M\Omega^{-1} \quad (9\text{–}24)$$

At this point it is convenient to introduce the parameters

$$x = \omega/K\sqrt{2RT_0}$$

and

$$y = \nu/K\sqrt{2RT_0}$$

and the complex probability integrals, Ω_N, which are defined as

$$\Omega_N(x,y) = i\pi^{-1}\int_{-\infty}^{\infty}\frac{dt\, e^{-t^2}t^N}{x - iy - t} \quad (9\text{–}25)$$

The Ω_0 have been tabulated (see the book by Faddeyva and Terent'ev cited in the References) and the Ω_N are thus known quantities too. The equation for $\bar{\bar{\rho}}$ can be rewritten in terms of these functions as

$$\bar{\bar{\rho}} = -y\bar{\bar{\rho}}\pi^{1/2}\Omega_0 - \frac{y\bar{\bar{u}}_z\pi^{1/2}\Omega_1}{\sqrt{RT_0/2}} - y\bar{\bar{T}}\pi^{1/2}(\Omega_2 - \tfrac{3}{2}\Omega_0) - \frac{S(K)\pi^{1/2}\Omega_0}{K\sqrt{2RT_0}} \quad (9\text{–}26)$$

where $\mathbf{K}$ has been taken along the z axis.

Equations similar to (9–26) can be obtained for $\bar{\bar{\mathbf{u}}}$ and $\bar{\bar{T}}$, and the resulting equations (there will be six altogether, since the equations for $\bar{\bar{\rho}}$, $\bar{\bar{\mathbf{u}}}$, and $\bar{\bar{T}}$ contain both a real and imaginary part) can be solved numerically. In Figure 9–2 we show some of the results that have been obtained.

9–5 SHOCK WAVE STRUCTURE

If the velocity in a particular flow field exceeds that of the local speed of sound, then it is possible that a shock wave may exist in the region where this occurs. Macroscopically a shock wave appears as a discontinuity across which there occur measurable changes in the thermofluid properties. Such phenomena are observed, for example, when a supersonic flow is turned, as in flow past a sharp corner which acts to

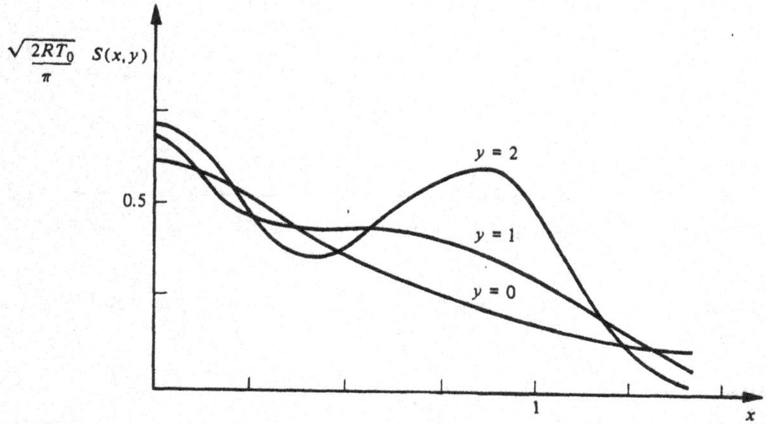

Figure 9–2 Response of a dilute gas to a density disturbance.

compress the flow, or flow over an airfoil, or some other body. Although the problem can be treated in an *ad hoc* manner using the macroscopic flow equations, the details concerning the properties of the flow through the shock wave cannot always be obtained from this point of view; often the shock must be treated as a pure discontinuity in the flow. This is in keeping with the spirit in which the conventional macroscopic flow equations are used. Shock waves which are microscopic in dimensions, are simply not amenable to a detailed treatment with these equations. (Note that this is not true, in principle, if the thirteen-moment equations are used. In this respect see Problem 9–9.) Thus we have a classic example of a problem which must be treated by the methods of kinetic theory. In approaching this problem we will depart from the procedure of the preceding two sections of using the model equation, and instead will use the Boltzmann equation itself. For such a strongly nonequilibrium problem it is clear that any linearized equation will be inadequate.

The specific problem which we will consider is that of determining the structure of a simple, steady shock wave in one dimension (see Figure 9–3). For this flow the Boltzmann equation reduces to the following form:

$$v_x \frac{\partial f}{\partial q_x} = J(f) \tag{9–27}$$

The boundary conditions which are used with Equation (9–27) are based on the state of the gas infinitely far from the shock wave, where the system is in equilibrium. If we use the subscript 1 to denote the thermofluid properties in the region behind the shock wave, for example, ρ_1, u_1, T_1, where these quantities do not depend on x, and the subscript 2 to serve

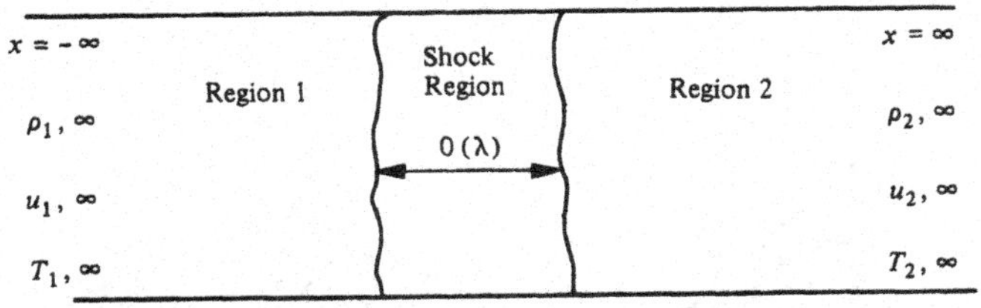

Figure 9–3 Flow configuration for steady one-dimensional shock wave.

the same purpose in the region ahead of the shock, the boundary conditions are then

$$\lim_{x \to -\infty} f(x,\mathbf{v}) = f_{\mathrm{LM}_1} = \frac{\rho_1}{(2\pi RT_1)^{3/2}} e^{-\frac{1}{2RT_1}((v_x-u_1)^2+v_y^2+v_z^2)}$$
$$\lim_{x \to \infty} f(x,\mathbf{v}) = f_{\mathrm{LM}_2} = \frac{\rho_2}{(2\pi RT_2)^{3/2}} e^{-\frac{1}{2RT_2}((v_x-u_2)^2+v_y^2+v_z^2)} \tag{9–28}$$

The net flow of mass, momentum, and energy into the shock must be equal to the net flow of these quantities out of the shock, so that we can relate the upstream and downstream values of the macroscopic variables through the following conservation equations:

$$\int d\mathbf{v}\, v_x f_{\mathrm{LM}_1} \begin{Bmatrix} 1 \\ v_x \\ v^2 \end{Bmatrix} = \int d\mathbf{v}\, v_x f_{\mathrm{LM}_2} \begin{Bmatrix} 1 \\ v_x \\ v^2 \end{Bmatrix} \tag{9–29}$$

Rewriting the above equations in terms of the thermo-fluid variables, we obtain the so-called Rankine–Hugoniot equations, which any solution we find must satisfy.

$$\rho_1 u_1 = \rho_2 u_2$$
$$\rho_1 u_1^2 + \rho_1 RT_1 = \rho_2 u_2^2 + \rho_2 RT_2$$
$$\rho_1 u_1 \left(\frac{u_1^2}{2} + \frac{5}{2} RT_1\right) = \rho_2 u_2 \left(\frac{u_2^2}{2} + \frac{5}{2} RT_2\right) \tag{9–30}$$

We have already stated that exact solutions of Boltzmann's equation are (with one exception) nonexistent, and thus it should come as no surprise that we will concern ourselves here in looking for an approximate solution to Equation (9–27). In particular, we will look for a superposition solution of the following form:

$$f(x,\mathbf{v}) = \frac{A_1(x)}{\rho_1} f_{\mathrm{LM}_1}(\mathbf{v}) + \frac{A_2(x)}{\rho_2} f_{\mathrm{LM}_2}(\mathbf{v}) \tag{9–31}$$

This particular approximation is originally due to Mott–Smith, who obtained the first solution of Boltzmann's equation for the shock structure problem in 1951. The particular form of the approximating function given by Equation (9–31) is called a bimodal distribution. More recent treatments of the shock structure problem involve introducing a third term on the right-hand side of Equation (9–31), or including additional structure into one of the terms in the bimodal distribution. However, the essential points of the solution technique are illustrated by considering the bimodal distribution as given in Equation (9–31), and we will accordingly restrict ourselves here to solutions of this form.

Only one independent equation for A_1 and A_2 can be obtained from the three conservation equations. To obtain a second independent equation we must therefore consider an additional moment equation. The simplest such equation is obtained by multiplying Boltzmann's equation (9–27) by $\int d\mathbf{v}\, v_x^2$ since the term $\int d\mathbf{v}\, v_x^2 J(f)$ has already been evaluated by us. This becomes evident when we write v_x^2 as

$$\begin{aligned} v_x^2 &= RTw_{0_x}^2 + 2w_{0_x}\sqrt{RT}\,u + u^2 \\ &= RT(H_{xx}^{(2)} + 1) + 2w_{0_x}\sqrt{RT}\,u + u^2 \end{aligned} \tag{9–32}$$

so that

$$\int d\mathbf{v}\, v_x^2 J(f) = \int d\mathbf{v}\, RTH_{xx}^{(2)} J(f) = -\beta\rho p_{xx} \tag{9–33}$$

where β and p_{xx} were defined in Chapter 7. The additional moment equation is then

$$\begin{aligned} \frac{\partial A_1}{\partial x}(3RT_1 + u_1^2) + \frac{\partial A_2}{\partial x}(3RT_2 + u_2^2) &= -\beta\rho p_{xx} \\ &= -\beta\tfrac{2}{3}A_1A_2(u_1 - u_2)^2 \end{aligned} \tag{9–34}$$

The final form of Equation (9–34) is obtained by writing p_{xx} as a function of f, and then substituting the bimodal form, Equation (9–31), for f.

We now have two independent equations for A_1 and A_2. These are Equation (9–34), and, from the mass conservation and first Rankine–Hugoniot equation,

$$\rho_1 u_1 = \rho_2 u_2 = \int d\mathbf{v}\, v_x f(x,\mathbf{v}) = A_1u_1 + A_2u_2 \tag{9–35}$$

Combining the above equation with Equation (9–34) to eliminate, for example, A_2, we then find

$$\frac{d}{dX}\left(\frac{A_1}{\rho_1}\right) = -\alpha\left[\frac{A_1}{\rho_1} - \left(\frac{A_1}{\rho_1}\right)^2\right] \tag{9–36}$$

Here the space variable x has been made dimensionless according to

$$X = \tfrac{2}{3}M\beta\frac{\rho_1}{u_1}x = \alpha x\lambda^{-1} \tag{9–37}$$

where the Mach number, $M = u_1(3/5RT_1)^{1/2}$, of the downstream fluid is the ratio of the downstream fluid velocity to the velocity of sound, and

$$\alpha = \frac{5}{2}\left(\frac{u_1}{u_2} - 1\right)^2 \left(\left[\frac{u_1}{u_2}\right]^2 - 1\right)^{-1} \tag{9-38}$$

For a hard sphere gas the collision integral is of order $\sigma^2 u_1$, and $\lambda \propto 1/\rho\sigma^2$; the corresponding relationship for general power law interparticle potentials, where $\sigma^2 u_1$ is replaced by β, has been used in Equation (9–37).

The solution to Equation (9–36) is

$$A_1(X) = \frac{\rho_1}{1 + e^{\alpha X}} \tag{9-39}$$

and from this result we can calculate the various profiles across the shock wave. The relative changes in the macroscopic quantities across the shock increase with increasing Mach number, for example,

$$\frac{\rho(X)}{\rho_1} = \frac{1 + \left(\dfrac{4M^2}{3 + M^2}\right) e^{\alpha X}}{1 + e^{\alpha X}} \tag{9-40}$$

In describing a shock wave the idea of a shock thickness proves useful. This quantity may be defined in terms of either ρ, $\mathbf{u}$, T as

$$\Delta\rho_i = \frac{\rho_{i_2} - \rho_{i_1}}{(\partial\rho_i/\partial X)_{\max}} \tag{9-41}$$

so that its numerical value for a given flow will depend on which of the ρ_i are used. All of the $\Delta\rho_i$ increase with M (a single exception to this rule is found in the case of hard sphere molecules, for which $\Delta\rho_i$ reaches a fixed limit as $M \to \infty$). In comparison with experimental results it is found that the Mott–Smith approximation is particularly good for describing strong shock waves (large Mach numbers), where more conventional theories tend to be exceptionally poor. For weak shock waves, however, the Mott–Smith results do not give particularly good agreement with experiment, and a more complicated form for f than a simple bimodal distribution must be used.

9–6 HALF-RANGE POLYNOMIAL EXPANSIONS

In the preceding section we considered an example of a flow where it appeared reasonable to approximate the distribution function by splitting it into two local Maxwellian-like distributions, each characteristic of a particular region of the physical space. The transition from one region to the other was continuous, as dictated by the physics of the flow. For a certain class of flow problems in kinetic theory the flow geometry dictates a similar splitting of the distribution function, this time depending on the region of velocity space which is being described. To see why such a procedure is desireable, consider the one-dimensional flow shown in Figure 9–4. The walls have different temperatures and

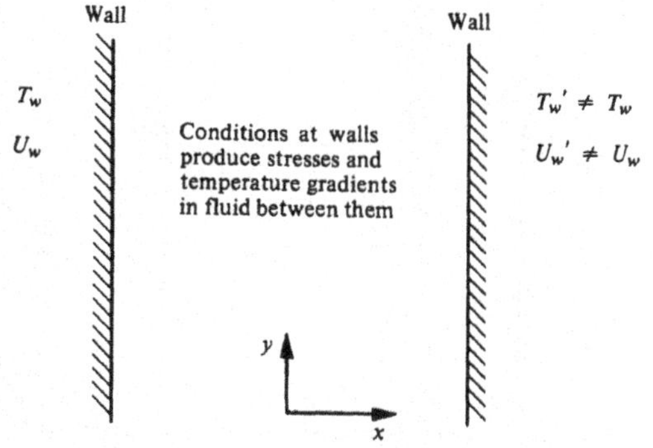

Figure 9–4 Flow configuration producing discontinuity in f at $v_x = 0$.

velocities, so that, for example, the distribution function in the vicinity of wall 2 for particles having $v_x > 0$ will be quite different than the distribution function for particles having $v_x < 0$. Thus we can anticipate that the distribution function which describes such flows will be discontinuous in the velocity space. As a result, the conventional approach to solving Boltzmann's equation for such flows, which is to expand f in polynomials defined over the full velocity space, has the definite shortcoming that many terms must be retained in the expansion to obtain a convergent result. Also, it is found that such full-range polynomial expressions for the distribution function only approximately satisfy the boundary conditions, which as we have seen are generally of a half-range nature.

The deficiencies of the full-range polynomial methods for the particular class of flows described above led Gross *et al.* to propose an expansion based on splitting the distribution function into two parts according to whether $v_x \lessgtr 0$, and separately expanding each of these half-range distribution functions. Examples of specific flows where this procedure is appropriate include the classical problems of one-dimensional heat flow between parallel plates, simple one-dimensional Couette flow, and the impulsive motion of a flat plate into an infinite fluid (Rayleigh's problem).

As an illustration of the half-range polynomial method we will consider the heat flow problem, reserving the Couette flow problem as an illustration for the discrete velocity model equation in the section that follows. Our treatment will follow that originally due to Gross and Ziering.

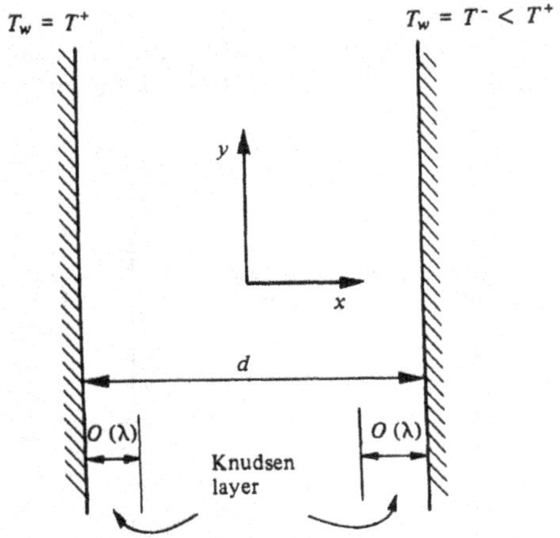

Figure 9–5 Geometry for simple one-dimensional heat flow.

The steady flow configuration to be studied is shown in Figure 9–5. The temperature gradient is assumed to be small,

$$\frac{T^+ - T^-}{T_0} \equiv \frac{\Delta T}{T_0} < 1$$

so that a linearized treatment is possible, and accordingly we will make use of the linearized Boltzmann equation. The linearization we adopt is

$$f = f_{LM}(x = 0)(1 + \phi) \tag{9–42}$$

where $f_{LM}(x = 0)$ is the Maxwellian distribution function characteristic of the plane $x = 0$,

$$f_{LM}(x = 0) = \frac{\rho_0}{(2\pi RT_0)^{3/2}} e^{-\mathbf{w}^2} \tag{9–43}$$

where the dimensionless velocity $\mathbf{w}$ is now given as $\mathbf{w} = \mathbf{v}(2RT_0)^{-1/2}$. The linearized Boltzmann equation then assumes the following form:

$$w_x \frac{\partial \phi}{\partial x} = (2RT_0)^{-1/2} L(\phi) \equiv \hat{L}(\phi) \tag{9–44}$$

This equation is to be solved subject to boundary conditions on f, which for simplicity we take as purely diffusive. Thus, if f^+, f^- denotes the distribution function for $v_x > 0$, $v_x < 0$, we will have

$$f^+\left(x = \frac{-d}{2}, \mathbf{v}\right) = \frac{A^+}{(2\pi RT^+)^{3/2}} e^{-\mathbf{v}^2/2RT^+} \qquad v_x > 0$$
$$f^-\left(x = \frac{d}{2}, \mathbf{v}\right) = \frac{A^-}{(2\pi RT^-)^{3/2}} e^{-\mathbf{v}^2/2RT^-} \qquad v_x < 0 \qquad (9\text{–}45)$$

where A^+ and A^- are to be determined. The boundary conditions on ϕ then follow from Equation (9–45) by linearization, and we find

$$\phi^+\left(x = \frac{-d}{2}, \mathbf{w}\right) = \frac{\Delta T}{T_0}(w^2 + B^+) \qquad w_x > 0$$
$$\phi^-\left(x = \frac{d}{2}, \mathbf{w}\right) = -\frac{\Delta T}{T_0}(w^2 - B^-) \qquad w_x < 0 \qquad (9\text{–}46)$$

where B^+ and B^- are the contribution of the A's.

We proceed by approximating ϕ by a sum of half-range polynomials in w. As we show in what follows, it is reasonable to use the following form for this approximation:

$$\phi \equiv \phi^+ = a_0{}^+ + a_1{}^+w_x + a_2{}^+w^2 + a_3{}^+w_xw^2 \qquad w_x > 0$$
$$\phi \equiv \phi^- = a_0{}^- + a_1{}^-w_x + a_2{}^-w^2 + a_3{}^-w_xw^2 \qquad w_x < 0 \qquad (9\text{–}47)$$

where the coefficients $a_i{}^+$ and $a_i{}^-$ are functions of space only. In general, $a_i{}^+ \neq a_i{}^-$; this is, of course, in keeping with the idea of having distinct distribution functions for opposite streams, which characterizes the half-range polynomial expansion method of solution. To show that Equation (9–47) is a reasonable form for ϕ, let us consider the form of the distribution function we would expect in the flow as based on the Chapman–Enskog theory. For the present flow, the second Chapman–Enskog approximation [see Equation (6–55)]

$$f_{\text{CE}} = f_{\text{LM}}(x)\left[1 + \frac{v_x}{(2RT)^{1/2}}\left(\frac{v^2}{2RT} - \frac{5}{2}\right)F(x)\right] \qquad (9\text{–}48)$$

where $F(x)$ is a function only of x. Linearizing according to Equation (9–42), we have $f_{\text{CE}} = f_{\text{LM}}(x = 0)(1 + \phi_{\text{CE}})$, and we find

$$\phi_{\text{CE}} = \bar{\rho} + \bar{T}(w^2 - \tfrac{3}{2}) + F(x)w_x(w^2 - \tfrac{5}{2}) \qquad (9\text{–}49)$$

where we have written

$$\begin{aligned} \rho(x) &= \rho(0)(1 + \bar{\rho}(x)) \\ T(x) &= T(0)(1 + \bar{T}(x)) \end{aligned} \qquad (9\text{-}50)$$

Since $F(x)$ includes the gradient $(\partial T/\partial x)$, it has also been considered as a pertubation term. In going from the full-range expansion of Equation (9–49) to a half-range expansion we must distinguish between ϕ for $v_x \gtrless 0$, and represent each part separately. Comparing Equations (9–49) and (9–47), we see that this is just what the latter representation accomplishes, and that it is the proper half-range generalization of the full-range expression.

Proceeding, we will consider the approximation for ϕ given by setting $a_1^{\pm}$, $a_3^{\pm} = 0$, so that we have

$$\begin{aligned} \phi^+ &= a_0^+ + a_2^+ w^2 \\ \phi^- &= a_0^- + a_2^- w^2 \end{aligned} \qquad (9\text{-}51)$$

Although the above approximation for ϕ is rather crude, it will serve here to illustrate how the a's are to be evaluated. Also, it can be shown that for $d/\lambda < 1$, that is, when collisions are not a primary effect, Equation (9–51) does give very good results (see Problem 9–8). In addition to the four unknowns which appear in Equation (9–51), we must add as unknowns the B's which appear in the boundary conditions, giving us a total of six unknowns to be determined. A simple physical argument which we will use again, allows us to equate B^+ and B^-, reducing the number of unknowns by one. This argument is based on the symmetry of the flow, which implies that

$$\begin{aligned} \phi^+(x,w_x) &= -\phi^-(-x,-w_x) \\ \phi^-(x,w_x) &= -\phi^+(-x,-w_x) \end{aligned} \qquad (9\text{-}52)$$

When this symmetry condition is applied to the boundary conditions, Equation (9–46), we find that $B^+ = B^- = B$. Another unknown can be eliminated by noting that there is an absence of flow between the two walls, that is, there is no accumulation of particles at either of the walls.

A set of four equations is now needed to determine the remaining four unknown quantities. These can be obtained by taking moments of the linearized Boltzmann equation. Before doing this let us rewrite Equation (9–51) as

$$\begin{aligned} \phi = \phi^+ + \phi^- &= \tfrac{1}{2}(a_0^+ + a_0^-) + \tfrac{1}{2}(a_2^+ + a_2^-)w^2 \\ &\quad + \tfrac{1}{2}[(a_0^+ - a_0^-) + (a_2^+ - a_2^-)w^2] \operatorname{sgn} w_x \end{aligned} \qquad (9\text{-}53)$$

where

$$\operatorname{sgn} w_x = +1 \qquad w_x > 0$$
$$= -1 \qquad w_x < 0 \tag{9–54}$$

By writing ϕ in the above form, and by electing to take moments with respect, for example, to $e^{-w^2}(1 + \operatorname{sgn} w_x)$, $e^{-w^2}(1 - \operatorname{sgn} w_x)$, $e^{-w^2}w^2(1 + \operatorname{sgn} w_x)$, $e^{-w^2}w^2(1 - \operatorname{sgn} w_x)$, the half-range character of the moment equations we will employ becomes especially clear. For example,

$$\int d\mathbf{w}\, e^{-w^2}(1 + \operatorname{sgn} w_x)\phi = \int_{-\infty}^{\infty} dw_z \int_{-\infty}^{\infty} dw_y \int_{-\infty}^{\infty} dw_x\, e^{-w^2}(1 + \operatorname{sgn} w_x)\phi$$
$$= 2\int_{-\infty}^{\infty} dw_z \int_{-\infty}^{\infty} dw_y \int_{0}^{\infty} dw_x\, e^{-w^2}\phi \tag{9–55}$$

and only ϕ^+ contributes to the above moment. Forming the indicated moment equations, we then obtain the following four equations:

$$\frac{d}{dx}(a_0^\pm + 2a_2^\pm) = (a_0^+ - a_0^-)\frac{I_1}{2\pi} + (a_2^+ - a_2^-)\frac{I_2}{2\pi} \tag{9–56}$$

$$\frac{d}{dx}(a_0^\pm + 3a_2^\pm) = (a_0^+ - a_0^-)\frac{I_2}{4\pi} + (a_2^+ - a_2^-)\frac{I_3}{4\pi} \tag{9–57}$$

where an obvious notational shorthand has been employed. The integrals I_i which appear are the contribution from the collision term, specifically,

$$I_1 = \int d\mathbf{w}\, e^{-w^2} \operatorname{sgn} w_x \hat{L}(\operatorname{sgn} w_x)$$
$$I_2 = \int d\mathbf{w}\, e^{-w^2} \operatorname{sgn} w_x \hat{L}(w^2 \operatorname{sgn} w_x)$$
$$= \int d\mathbf{w}\, e^{-w^2} w^2 \operatorname{sgn} w_x \hat{L}(\operatorname{sgn} w_x)$$
$$I_3 = \int d\mathbf{w}\, e^{-w^2} w^2 \operatorname{sgn} w_x \hat{L}(w^2 \operatorname{sgn} w_x) \tag{9–58}$$

These integrals have been evaluated for both hard spheres and Maxwell molecules; the results are, respectively,

$$I_1 = -0.586\frac{\pi}{\lambda}, \quad -1.337\frac{\pi}{\lambda}$$

$$I_2 = -0.672\frac{\pi}{\lambda}, \quad -1.482\frac{\pi}{\lambda}$$

$$I_3 = -2.547\frac{\pi}{\lambda}, \quad -4.202\frac{\pi}{\lambda} \tag{9–59}$$

where the mean free path for Maxwell molecules is

$$\lambda = 2\rho\pi A_2(5)(K/RT)^{1/2}$$

The set of coefficient equations (9–56), (9–57) can be solved directly, and we find

$$a_i^\pm = A_i^\pm + C_i^\pm x \qquad i = 0, 1 \tag{9–60}$$

The constant coefficients $A_i^{\pm}$, $C_i^{\pm}$ can be evaluated by again making use of the symmetry properties of the flow, which imply

$$\begin{aligned} a_0^+(x) &= -a_0^-(-x) \\ a_2^+(x) &= -a_2^-(-x) \end{aligned} \tag{9–61}$$

and the condition of zero mass flow, which we can write in terms of the a's as

$$a_0^+ - a_0^- + 2(a_2^+ - a_2^-) = 0 \tag{9–62}$$

The above equations, plus the boundary conditions on the a's [see Equations (9–46) and (9–47)],

$$\begin{aligned} a_0^{\pm}\left(\mp \frac{d}{2}\right) &= \pm \frac{\Delta T}{T_0} B \\ a_2^{\pm}\left(\mp \frac{d}{2}\right) &= \pm \frac{\Delta T}{T_0} \end{aligned} \tag{9–63}$$

serve to fully determine the solution, and we find, after combining the above equations,

$$\begin{aligned} a_0^{\pm} &= A_0^+\left[\pm 1 + \left(3I_1 - \tfrac{5}{2}I_2 + \frac{I_3}{2}\right)\frac{x}{\pi}\right] \\ a_2^{\pm} &= A_0^+\left[\mp \frac{1}{2} + \left(I_2 - \frac{I_3}{4} - I_1\right)\frac{x}{\pi}\right] \end{aligned} \tag{9–64}$$

with

$$A_0^+ = -2\frac{\Delta T}{T_0}\left(\frac{1}{1 + \frac{d}{\pi}\left[I_2 - \frac{I_3}{4} - I_1\right]}\right) \tag{9–65}$$

The physical quantities of interest are the density and temperature profiles between the walls. A simple calculation shows that

$$\begin{aligned} \rho &= \frac{\rho(0)}{2}\left[a_0^+ + a_0^- + \tfrac{3}{2}(a_2^+ + a_2^-)\right] \\ T &= T(0)\left[1 + \left(\frac{a_2^+ + a_2^-}{2}\right)\right] \end{aligned} \tag{9–66}$$

so that in the present approximation we find

$$\begin{aligned} \rho &= \rho(0)\left[1 + A_0^+\frac{x}{2\pi}\left(\tfrac{3}{2}I_1 - I_2 + \frac{I_3}{8}\right)\right] \\ T &= T(0)\left[1 + A_0^+\frac{x}{2\pi}\left(I_2 - \frac{I_3}{4} - I_1\right)\right] \end{aligned} \tag{9–67}$$

The above expressions can be simplified by substituting for the I_i, and we find, for example, for hard spheres

$$\rho = \rho(0)\left\{1 + \frac{\left(1.352\,\frac{x}{\lambda}\right)}{\left(1 + 0.551\,\frac{d}{\lambda}\right)}\frac{\Delta T}{T_0}\right\} \tag{9-68}$$

$$T = T(0)\left\{1 - \frac{\left(0.551\,\frac{x}{\lambda}\right)}{\left(1 + 0.551\,\frac{d}{\lambda}\right)}\frac{\Delta T}{T_0}\right\}$$

As mentioned above, the above results are very good for $d/\lambda < 1$, but give only a crude approximation for significantly smaller mean free paths. For this regime the full form of ϕ as given by Equation (9–47) must be retained. The method of solution proceeds as above, except now additional moments must be taken to generate equations to determine the extra coefficients which appear. These results display considerably more structure than those given here. In particular, the Knudsen layer, which is the region within several mean free paths of the walls, is given to a high degree of accuracy.

9–7 COUETTE FLOW

As a final example we consider the steady, one-dimensional Couette flow problem in the context of the discrete velocity model. Although this model is very crude, surprisingly accurate qualitative results are obtained. Specifically, we will use an eight-velocity, one-speed model to furnish the basic molecular description, with the velocities taken in the directions defined by the interior diagonals of the unit cube, as shown in Figure 9–6. There will be two types of collision which must be considered as being effective, consistent with momentum conservation, in changing the number of particles moving in the various directions. These are head-on collisions, and collisions of the type

$$\mathbf{v}_1, \mathbf{v}_4 \xrightarrow[\text{collision}]{\text{binary}} \begin{matrix}\mathbf{v}_1, \mathbf{v}_4 \\ \mathbf{v}_2, \mathbf{v}_3\end{matrix}$$

$$\mathbf{v}_2, \mathbf{v}_8 \xrightarrow[\text{collision}]{\text{binary}} \begin{matrix}\mathbf{v}_2, \mathbf{v}_8 \\ \mathbf{v}_4, \mathbf{v}_6\end{matrix} \tag{9-69}$$

which involve velocities corresponding to opposite points on the diagonals of the six faces of the unit cube.

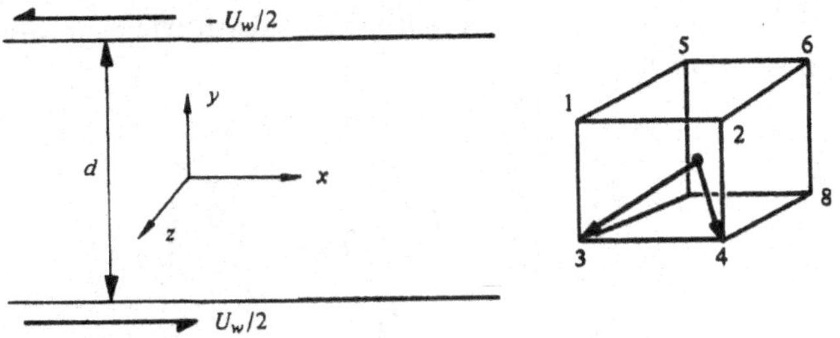

Figure 9–6 Flow geometry with velocity space for Covette flow problem. The eight allowable molecular velocities are specified by the interior diagonals of the cube, and labeled as shown. The v_3 and v_4 directions are indicated in the figure.

Let σ be the mutual collision cross section, $v^* = \sqrt{3}\, v$ the particle speed (so that v is the component in the x, y, and z directions), N_i the number of particles per unit volume with velocity in the i direction, $n = \sum_i N_i$, and $n_i = N_i/n$. In the one-dimensional flow problem under consideration the flow is symmetric about the x-y plane so that $n_i = n_{i+4}$, $i = 1 - 4$. In the equilibrium state all the n_i are equal. The collision rate for a particle moving in a given direction, for example, the one direction, in this equilibrium state is then

$$\begin{aligned}\nu_1 &= v^*\sigma\Big[\frac{2}{\sqrt{3}}\{N_1N_2 + N_1N_3 + N_1N_5\} \\ &\quad + 2\sqrt{\tfrac{2}{3}}\,\{N_1N_4 + N_1N_6 + N_1N_7\} \\ &\quad + 2N_1N_8\Big] \\ &= 10.37 v^*\sigma N_1^2\end{aligned} \tag{9–70}$$

and the collision frequency per particle is then

$$\nu = \frac{\nu_1}{N_1} = \frac{10.37}{8}\, v^*\sigma n \tag{9–71}$$

Accordingly, the mean free path is

$$\lambda = \nu v^{*-1} = \frac{8}{10.37\sigma n}$$

The model Boltzmann equation is given by a set of four simultaneous equations of the form

$$\frac{\partial N_i}{\partial t} + \mathbf{v}_i^* \cdot \frac{\partial N_i}{\partial \mathbf{q}} = G_i - L_i \tag{9–72}$$

where G_i and L_i denote the gain and loss of the n_i per unit time due to binary collisions. Let us consider $i = 1$; we have [see Equation (8–23)]

$$\frac{\partial N_1}{\partial t} - v\frac{\partial N_1}{\partial x} + v\frac{\partial N_1}{\partial y} = G_1 - L_1$$

$$= \left[2\sqrt{\tfrac{2}{3}}\,v^*\frac{\sigma}{2}(N_2N_3 + N_2N_5 + N_3N_5) + 2v^*\frac{\sigma}{4}(N_2N_7 + N_3N_6 + N_4N_5)\right]$$

$$- \left[2\sqrt{\tfrac{2}{3}}\,v^*\frac{\sigma}{2}(N_1N_4 + N_1N_6 + N_1N_7) + 2v^*\sigma\tfrac{3}{4}N_1N_8\right] \qquad (9\text{–}73)$$

Setting

$$\theta = \tfrac{1}{2}(1 + \sqrt{\tfrac{2}{3}})v^*\sigma n = 0.7v^*\lambda^{-1}$$

and taking into account the symmetry conditions, we can rewrite Equation (9–73) as

$$\frac{\partial n_1}{\partial t} - v\frac{\partial n_1}{\partial x} + v\frac{\partial n_1}{\partial y} = 2\theta(n_2n_3 - n_1n_4) \qquad (9\text{–}74)$$

The remaining equations, which are similarly determined, are

$$\frac{\partial n_2}{\partial t} + v\frac{\partial n_2}{\partial x} + v\frac{\partial n_2}{\partial y} = 2\theta(n_1n_4 - n_2n_3)$$

$$\frac{\partial n_3}{\partial t} - v\frac{\partial n_3}{dx} - v\frac{\partial n_3}{\partial y} = 2\theta(n_1n_4 - n_2n_3)$$

$$\frac{\partial n_4}{\partial t} + v\frac{\partial n_4}{\partial x} - v\frac{\partial n_4}{\partial y} = 2\theta(n_2n_3 - n_1n_4) \qquad (9\text{–}74')$$

The symmetry of the flow dictates that there be no net flow in either the y or z directions, and thus

$$u_y = \sum_i n_i v_{iy}^* = 0 \qquad (9\text{–}75)$$

from which it follows that $n_1 + n_2 = n_3 + n_4 = \frac{1}{4}$. Using this relationship to eliminate n_2 and n_4 from Equation (9–74) and taking into account that in the flow under consideration the n_i are independent of both x and t, the model Boltzmann equation can be reduced to the following two equations:

$$v\frac{\partial n_1}{\partial y} = \frac{\theta}{2}(n_3 - n_1)$$

$$-v\frac{\partial n_3}{\partial y} = \frac{\theta}{2}(n_1 - n_3) \qquad (9\text{–}76)$$

It will again prove convenient to use dimensionless variables, in this case $Y = y/d$ and $\alpha = \theta d/2v$, in terms of which the above equations become

$$\frac{\partial n_1}{\partial Y} = \alpha(n_3 - n_1) = \frac{\partial n_3}{\partial Y} \tag{9-77}$$

This equation must be supplemented by boundary conditions, which specify the flow at each of the walls. We will again adopt the conditions of pure diffuse reflection, which imply that the average velocity of the reflected particles at a wall are equal to the wall velocity, so that

$$v[n_2(-\tfrac{1}{2}) - n_1(-\tfrac{1}{2})][n_1(-\tfrac{1}{2}) + n_2(-\tfrac{1}{2})]^{-1} = \frac{U_w}{2}$$

$$v[n_4(\tfrac{1}{2}) - n_3(\tfrac{1}{2})][n_3(\tfrac{1}{2}) + n_4(\tfrac{1}{2})]^{-1} = \frac{-U_w}{2} \tag{9-78}$$

or, using Equation (9–75),

$$n_1(-\tfrac{1}{2}) = \frac{1}{8}\left(1 - \frac{U_w}{2v}\right)$$

$$n_3(\tfrac{1}{2}) = \frac{1}{8}\left(1 + \frac{U_w}{2v}\right) \tag{9-79}$$

The solutions to Equation (9–77) which satisfy the above boundary conditions are

$$n_1 = \frac{1}{8}\left[\left(\frac{\alpha}{\alpha + 1}\right)\frac{U_w}{v} Y + 1 - \frac{1}{2(\alpha + 1)}\frac{U_w}{v}\right]$$

$$n_3 = \frac{1}{8}\left[\left(\frac{\alpha}{\alpha + 1}\right)\frac{U_w}{v} Y + 1 - \frac{1}{2(\alpha + 1)}\frac{U_w}{v}\right] \tag{9-80}$$

Using these results we can directly calculate the velocity field

$$u_x = \sum_i n_i v_{ix}^*$$

and the shear tensor

$$P_{xy} = \rho \sum_i n_i v_{ix}^* v_{iy}^*$$

and we find

$$u_x = -\frac{\alpha}{\alpha + 1} U_w Y \tag{9-81}$$

$$P_{xy} = \frac{\rho v U_w}{2(\alpha + 1)} \tag{9-82}$$

The above result can be used to determine a viscosity coefficient for the model, since

$$\mu = P_{yz}\left(\frac{\partial U_z}{\partial y}\right)^{-1} = 0.48\rho v^*\lambda \tag{9–83}$$

which, with v^* identified with the mean thermal speed, gives close agreement with the values found in Chapters 6 and 7.

The result obtained for the velocity field exhibits the so-called velocity slip effect, which is characteristic of the Knudsen layer. Note first that $\alpha \propto d/\lambda$. Therefore if we define the fractional slip velocity as

$$\frac{2\,\Delta U_z}{U_w} = \frac{\frac{1}{2}U_w - u_z(-\frac{1}{2})}{\frac{1}{2}U_w} \tag{9–84}$$

we will then have

$$\frac{2\,\Delta U_z}{U_w} = \frac{1}{1+\alpha} \tag{9–85}$$

which qualitatively reproduces results obtained using far more sophisticated methods. The above result correctly predicts that the velocity of the fluid in the layer adjacent to the walls is not equal to the wall velocity; instead there is a slipping of the fluid at the wall. The magnitude of this velocity slip effect depends on the fluid properties and the flow geometry through the parameter d/λ. For highly continuum flow, $\alpha \rightarrow \infty$, and the slip effect vanishes; but for flows at finite α there will be a definite effect.

References

Rigorous existence proofs for Boltzmann's equation have been given by:

1. T. Carleman (1957). See General References.
2. H. Grad, *Proc. Am. Math. Soc., Symp. in Appl. Math.*, vol. 20. Providence, R.I.: American Mathematical Society, 1969.
3. D. Morgenstern, *Proc. Natl. Acad. Sci.*, vol. 40, p. 719, 1954.
4. ——, *J. Rational Mech. Anal.*, vol. 4, p. 533, 1955.
5. E. Wild, *Proc. Cambridge Phil. Soc.*, vol. 47, p. 602, 1951.

Boundary conditions for flow problems in kinetic theory have been discussed by:

6. H. Grad (1949). See References, Chapter 7.
7. —— (1958). See General References.
8. J. C. Maxwell (1965). See General References.

The plane Poiseuille flow problem has been considered by:

9. C. Cercignani, in *Rarefied Gas Dynamics*, J. A. Laurmann, Ed. New York: Academic Press, 1963.

10. ——, *J. Math. Anal. Appl.*, vol. 12, p. 234, 1965.

11. M. Knudsen, *Ann. Physik*, vol. 28, p. 75, 1909.

12. S. Ziering, in *Rarefied Gas Dynamics*, L. Talbot, Ed. New York: Academic Press, 1961.

The Abramowitz integrals are considered in considerable detail, and asymptotic values for them are obtained by:

13. M. Abramowitz, *J. Math. Phys.*, vol. 37, p. 188, 1953.

The van Hove correlation function was proposed by:

14. L. van Hove, *Phys. Rev.*, vol. 95, p. 249, 1954.

Boltzmann's equation has been used to calculate $G(\mathbf{q},t)$ by:

15. S. Ranganathan and S. Yip, *Phys. Fluids*, vol. 9, p. 372, 1966.

16. A. Sugawara, S. Yip, and L. Sirovich, *Phys. Fluids*, vol. 11, p. 925, 1968.

17. S. Yip and M. Nelkin, *Phys. Rev.*, vol. 135, p. A1241, 1964.

The complex probability integrals which appear in Section 9–4 are tabulated in:

18. V. N. Faddeyva and N. M. Terent'ev, *Tables of the Values of the Function* $w(z) = e^{-z^2}\left(1 + (2i/\sqrt{\pi}) \int_0^z dt\, e^{t^2}\right)$. New York: Pergamon, 1961.

The bimodal approximation was first proposed, in connection with the shock wave problem, by:

19. H. M. Mott-Smith, *Phys. Rev.*, vol. 82, p. 885, 1951.

The shock wave problem has also been treated by the thirteen-moment method by:

20. H. Grad, *Commun. Pure and Appl. Math.*, vol. 5, p. 257, 1952.

21. L. W. Holway, Jr., *Phys. Fluids*, vol. 7, p. 911, 1964.

Generalizations of the Mott-Smith method have been proposed by:

22. P. Glansdorff, in *Rarefied Gas Dynamics*, L. Talbot, Ed. New York: Academic Press, 1961.

23. H. Salwen, C. Grosch, and S. Ziering, *Phys. Fluids*, vol. 7, p. 180, 1964.

24. S. Ziering, F. Ek, and P. Koch, *Phys. Fluids*, vol. 4, p. 975, 1961.

The half-range polynomial method was introduced by:

25. E. P. Gross, E. A. Jackson, and S. Ziering, *Ann. Phys.*, vol. 1, p. 141, 1957.

Applications of the half-range method have been considered by:

26. E. P. Gross and E. A. Jackson, *Phys. Fluids*, vol. 2, p. 432, 1959.

27. ——, *Phys. Fluids*, vol. 1, p. 318, 1958.

28. E. P. Gross and S. Ziering, *Phys. Fluids*, vol. 1, p. 215, 1958.

29. ——, *Phys. Fluids*, vol. 2, p. 701, 1959.

30. S. Ziering, *Phys. Fluids*, vol. 3, p. 503, 1960.

The last two references listed above treat the heat flow problem discussed in the test. The Couette flow problem has been studied in the context of the discrete velocity model by:

31. J. Broadwell, *J. Fluid Mech.*, vol. 19, p. 401, 1964.

This problem has also been considered in References 25, 28, and 30 by means of the half-range method.

Many of the flow problems discussed in this chapter, as well as a number of other flow problems are discussed in

32. M. N. Kogan, *Rarefied Gas Dynamics*. New York: Plenum Press, 1969.

The above list of references is far from being all inclusive; the quantity of the relevant literature precludes any efforts in this direction. Many other articles, as well as additional references, can be found in the *Proceedings of the International Symposium in Rarefied Gas Dynamics*, which is a continuing series.

The exact solution to Boltzmann's equation mentioned in the text can be found in Reference 5 of Chapter 7.

Problems

9–1. Show that the quantity α which appears in Equation (9–3) as the accommodation coefficient, that is, the fraction of specularly reflected particles at the boundary, can also be related to the fraction of the total tangential momentum transferred to the boundary by incident molecules, specifically that this fraction is just $(1 - \alpha)$.

9–2. Prove the relationship for the Abramowitz integrals given in Equation (9–17).

9–3. Show that $J_0(z)$ satisfies the differential equation $zJ_0''' + J_0'' + 2J_0 = 0$ with $J_0(0) = \pi^{1/2}/2$.

9–4. For large values of its argument

$$J_0(\tau) \to \frac{\pi}{3} e^{1/2 - 3\sqrt{z/2}}(1 + \tfrac{5}{36}z^{-1/2})$$

Use this to show that at high pressure, that is, $\tau^{-1} \gg 1$, Equation (9–18) gives the classical Poiseuille solution, $Q/Ad^2 = \text{constant}$.

9–5. Find $G(\mathbf{q},t)$ for a highly rarefied gas by solving the Boltzmann equation with the collision term set equal to zero, subject to the initial condition

$$f(\mathbf{q},\mathbf{v},0) = \frac{\rho_0 e^{-\mathbf{v}^2/2RT_0}}{(2\pi RT_0)^{3/2}} \delta(\mathbf{q})$$

9–6. Suppose we attempt to solve the shock problem discussed in Section 9–5 by using the trimodal distribution function

$$f(x,\mathbf{v}) = \frac{A_1(x)}{\rho_1} f_{\mathrm{LM}_1}(\mathbf{v}) + \frac{A_2(x)}{\rho_2} f_{\mathrm{LM}_2}(\mathbf{v}) + B_3(x)f_3(\mathbf{v})$$

where the first two terms on the right-hand side are the bimodal distribution given by Equation (9–31), and

$$f_3(\mathbf{v}) = \frac{1}{(\pi RT_3)^{3/2}} (v_x - u_3)e^{-\frac{1}{RT_3}[(v_x-u_3)^2+v_y^2+v_z^2]}$$

with u_3 and T_3 constants. What are the values of $B_3(\pm\infty)$? Determine u_3 and T_3 by making use of the Rankine–Hugoniot equations and the moment equations calculated with the trimodal distribution function.

9–7. Suppose we wished to solve the shock problem of Section 9–5 by considering an expansion of f in Hermite polynomials. We could show that the solution obtained would only be convergent for $T_2 \leq 2T_1$. What upper limit does this impose on M?

9–8. Determine the distribution function which describes the flow studied in Section 9–6 for a highly rarefied gas (so that the collision term can be set equal to zero) and explicitly demonstrate its half-range character.

9–9. Verify Equations (9–56) and (9–57) by forming the appropriate moments of the linearized Boltzmann equation.

9–10. The Couette flow problem discussed in Section 9–7 can also be studied in terms of a two-dimensional discrete velocity model having four properly chosen velocities. Rederive the results found in Section 9–7 using this two-dimensional model.

9–11. The impulsive motion of an infinite plane bounding a semi-infinite region of gas, which is initially in equilibrium, is one of the classical problems of fluid physics (the Rayleigh problem). Solve this problem (that is, determine the n_i) for the case where the plane $y = 0$ moves with velocity $U_w \ll v$ in the x direction, using the discrete velocity model of Section 9–7. Assume diffusive boundary conditions.

CHAPTER 10

gas mixtures

10–1 DESCRIPTION OF BINARY MIXTURES

Our discussion so far has been limited to systems of identical particles. In the present chapter we will consider the necessary modifications which must be made to extend our earlier results to systems composed of more than one molecular species. In particular, we will focus our attention on two-component systems. Since the basic molecular mechanism of the Boltzmann equation description is binary collisions, two-component systems will exhibit all of the complexities of n-component systems. The generalization of the results we will obtain to systems containing more than two species is straightforward, and we will therefore not bother to consider such systems except in the problems at the end of the chapter.

It is possible to reduce the Liouville equation for a two-component system in much the same manner as was done for a pure system. Prescribing the appropriate BGL, we then obtain two simultaneous equations for the distribution functions $f_A(\mathbf{q},\mathbf{v},t)$ and $f_B(\mathbf{q},\mathbf{v},t)$ describing the two species, which we will call A and B. (It will be most convenient in this chapter to use number densities rather than mass densities, and since the f_i will always have a subscript attached to them we will not bother to add the n subscript also, but rather we adopt the convention that in this chapter f_i denotes a number density.) The simultaneous Boltzmann equations are

$$\frac{\partial f_A}{\partial t} + \mathbf{v} \cdot \frac{\partial f_A}{\partial \mathbf{q}} = J_{AA}(f) + J_{AB}(f) \tag{10–1}$$

$$\frac{\partial f_B}{\partial t} + \mathbf{v} \cdot \frac{\partial f_B}{\partial \mathbf{q}} = J_{BB}(f) + J_{BA}(f) \tag{10–2}$$

In general, for a k-component system, there will be k such equations, each containing k terms on the right-hand side.

The appearance of two collision terms on the right-hand side of Equations (10–1) and (10–2) is a result of the possibility for the f_i to change through both self collisions (A–A collisions and B–B collisions) or cross collisions (A–B collisions). The general collision term is

$$J_{ij}(f_1) = \int d\mathbf{v}_2\, d\theta\, d\epsilon\, B_{ij}(\theta,V)[f_{i1}'f_{j2}' - f_{i1}f_{j2}] \qquad i = A, B;\, j = A, B \tag{10–3}$$

Note that three distinct B_{ij}'s will appear in Equations (10–1) and (10–2), B_{AA}, which describes the self collisions between two A particles, B_{BB}, and $B_{AB} = B_{BA}$, which are similarly defined. Further, in order to keep our notation as uncluttered as possible, we have not distinguished between the $\mathbf{v}_1'$, $\mathbf{v}_2'$ (which the notation used in Equation (10–3) implies appear in the arguments of the f_i') resulting from an A–A, B–B, or A–B collision; but it should be kept in mind that these post-collisional velocities will depend on the identity of the colliding particles. In practice, it will always be clear, for example, from the particular B_{ij} which also appears in the collision integral, which of the three possible collisions are involved. The reason that these distinctions are necessary is that the interparticle potentials which describe the various types of mutual interactions in a mixture are not, in general, identical.

The moments of the distribution functions f_A and f_B can be related to the observable macroscopic properties in much the same way as was done for a pure system. Thus (remember, in this chapter the f_i are number densities)

$$n_i = \int d\mathbf{v}\, f_i \tag{10–4}$$

is the macroscopic number density of species i, whereas

$$n = \sum_{i=A,B} n_i \tag{10–5}$$

is the macroscopic number density of the composite system. The mass density of species i is

$$\rho_i = \int d\mathbf{v}\, m_i f_i = m_i n_i \tag{10–6}$$

and for the composite system

$$\rho = \sum_{i=A,B} \rho_i \tag{10–7}$$

The mass flow velocity $\mathbf{u}$ is defined as the weighted average

$$\rho\mathbf{u} = \sum_{i=A,B} \int d\mathbf{v}\, m_i \mathbf{v} f_i \equiv \sum_{i=A,B} \rho_i \mathbf{u}_i \tag{10–8}$$

We will again find it convenient to use the peculiar velocity $\mathbf{v}_0 = \mathbf{v} - \mathbf{u}$, and in addition, we will also make use of the diffusion velocities $\mathbf{v}_{Di} = \mathbf{u}_i - \mathbf{u}$. The latter quantities, which indicate the rate of flow of the ith species with respect to the composite system, can also be defined as averages of the peculiar velocity. The pressure tensor of the individual species is

$$\mathsf{P}_i = \int d\mathbf{v}\; m_i \mathbf{v}_0 \mathbf{v}_0 f_i \tag{10–9}$$

and the pressure tensor for the composite system is the sum of the P_i,

$$\mathsf{P} = \sum_{i=A,B} \mathsf{P}_i \tag{10–10}$$

The temperature can be most simply defined by making use of Dalton's law,

$$p = \sum_{i=A,B} \rho_i R_i T$$

so that

$$p = \tfrac{1}{3}\mathsf{P} : \mathsf{U} = nkT \tag{10–11}$$

(remember that Boltzmann's constant $k = R_i/m_i$) or

$$3nkT = \sum_{i=A,B} \int d\mathbf{v}\; m_i v_0^2 f_i \tag{10–12}$$

The equations of motion for n_i, ρ, $\mathbf{u}$, and T are found by taking moments of the simultaneous Boltzmann equations. Multiplying these equations by 1 and integrating, we find

$$\frac{\partial n_i}{\partial t} + \frac{\partial}{\partial \mathbf{q}} \cdot n_i \mathbf{u}_i = 0 \tag{10–13}$$

Multiplying this equation by m_i, and adding the equations for species A and B, we obtain the usual equation of continuity:

$$\frac{\partial \rho}{\partial t} + \frac{\partial}{\partial \mathbf{q}} \cdot \rho \mathbf{u} = 0 \tag{10–14}$$

The momentum equation for the individual species includes a contribution from the collision term, expressing the fact that it is the total momentum of the system rather than the species momentum which is a summational invariant. This contribution to the species equations does not appear in the density equations since all particles, regardless of their identity, are conserved in a collision (unless the possibility of a chemical reaction is considered). Proceeding directly to the momentum equation for the composite system, we multiply the Boltzmann equations by $m_i\mathbf{v}_0$, integrate, and add the resulting equations to obtain

$$\rho \frac{D\mathbf{u}}{Dt} + \frac{\partial}{\partial \mathbf{q}} \cdot \mathsf{P} = 0 \tag{10–15}$$

The kinetic energy, like the momentum, is only conserved for the mixture as a whole. Multiplying the Boltzmann equations by $m_i v_0^2$, integrating, adding the resulting equations, and making use of the equipartition theorem, we obtain

$$\tfrac{3}{2}n \frac{D(kT)}{Dt} + \frac{\partial}{\partial \mathbf{q}} \cdot \mathbf{Q} - \tfrac{3}{2}kT \frac{\partial}{\partial \mathbf{q}} \cdot \sum_{i=A,B} n_i \mathbf{v}_{Di} + \mathsf{P} : \frac{\partial \mathbf{u}}{\partial \mathbf{q}} = 0 \tag{10–16}$$

where the heat flux vector $\mathbf{Q}$ is the sum of the species heat flux vectors, and these are defined in the usual way.

If we define the H function for a mixture as

$$H = \sum_{i=A,B} H_i = \sum_i \int d\mathbf{v}\, f_i \ln f_i \tag{10–17}$$

then it can be shown that the H function decays monatonely to a unique equilibrium state which is characterized by the distributions

$$f_{Mi}(\mathbf{v}) = \frac{n_i e^{-v_0^2/2R_iT}}{(2\pi R_i T)^{3/2}} \tag{10–18}$$

In certain cases we can distinguish certain regimes which will characterize the approach to equilibrium. Of special interest is the case where one of the species is significantly heavier than the other. In this case the system will evolve from its initial nonequilibrium state to a state where first the light species, then the heavy species, and finally the composite system itself equilibrates.

10–2 TRANSPORT PHENOMENA IN MIXTURES

All the methods of solution developed for treating Boltzmann's equation have been applied to the equations for a gas mixture. Although there are certain differences in detail, the methods themselves can be used almost without modification. In this section, we will use the Chapman–Enskog procedure to obtain the closed form of the transport equations at the Euler and Navier–Stokes levels of description, corresponding to the first and second Chapman–Enskog approximations.

Proceeding as for a pure system, we expand the f_i in the small parameter δ, and insert an δ^{-1} in front of each of the collision terms in Equations (10–1) and (10–2). In the first approximation the f_i are given by local Maxwellian distributions. The thermo-fluid equations in this approximation are the Euler equations for a mixture, which are characterized by the absence of transport processes. In addition to the flow of heat and momentum, we must now also consider species transport, or diffusion. We can easily show that none of the above mentioned transport processes appear in the Euler equations by calculating $\mathbf{Q}^{(0)} = 0$, $\mathbf{P}^{(0)} = p\boldsymbol{\delta}$, $\mathbf{u}_i = \mathbf{u}$. The last relationship implies that the diffusion velocities will be zero. In general, the diffusion velocities need not be zero, since the constraints on the distribution functions are taken as

$$\int d\mathbf{v}\, f_i^{(j)} = n_i\delta_{0j}$$

$$\sum_{i=A,B} \int d\mathbf{v}\, f_i^{(j)} m_i\mathbf{v} = \rho\mathbf{u}\delta_{0j}$$

and

$$\sum_{i=A,B} \int d\mathbf{v}\, f_i^{(j)} m_i v_0^2 = 3nkT\delta_{0j}$$

In the second Chapman–Enskog approximation we set $f_i = f_{\mathrm{LM}i}(1 + \phi_i)$. The time derivative which appears on the left-hand side of the Boltzmann equations is replaced by

$$\frac{\partial f_i}{\partial t} \longrightarrow \frac{\partial f_{\mathrm{LM}i}}{\partial n_i}\frac{\partial_0 n_i}{\partial t} + \frac{\partial f_{\mathrm{LM}i}}{\partial \mathbf{u}}\cdot\frac{\partial_0 \mathbf{u}}{\partial t} + \frac{\partial f_{\mathrm{LM}i}}{\partial T}\frac{\partial_0 T}{\partial t} \tag{10–19}$$

and the space derivative by $(\partial f_{\mathrm{LM}i}/\partial \mathbf{q})$. The sum of these terms is again denoted as $(D_0 f_i/Dt)$. The simultaneous equations for the ϕ_i are then

$$\frac{D_0 f_A}{Dt} = n_A{}^2 I_{AA}(\phi_A) + n_A n_B I_{AB}(\phi_A + \phi_B)$$

$$\frac{D_0 f_B}{Dt} = n_B{}^2 I_{BB}(\phi_B) + n_B n_A I_{BA}(\phi_B + \phi_A) \tag{10–20}$$

where the I_{ij} are linear operators similar to the I operator introduced in Equation (6–44). Evaluating the left-hand sides of Equations (10–20) we obtain

$$\frac{D_0 f_A}{Dt} = f_{\mathrm{LM}A}\left\{\left(\frac{v_0{}^2}{2R_A T} - \frac{5}{2}\right)\mathbf{v}_0\cdot\frac{\partial \ln T}{\partial \mathbf{q}} + \frac{1}{R_A T}(v_{0\alpha}v_{0\beta} - v_0{}^2\delta_{\alpha\beta})\frac{\partial u_\beta}{\partial q_\alpha} + \frac{n}{n_A}\mathbf{v}_0\cdot\mathbf{d}_A\right\} \tag{10–21}$$

and a similar equation for $(D_0 f_B/Dt)$. The vector $\mathbf{d}_A$ which occurs in the above equation is

$$\mathbf{d}_A = \frac{\partial}{\partial \mathbf{q}}\left(\frac{n_A}{n}\right) + \left(\frac{n_A}{n} - \frac{\rho_A}{\rho}\right)\frac{\partial \ln p}{\partial \mathbf{q}} = -\mathbf{d}_B \tag{10-22}$$

The above results allow us to infer that the nonequilibrium component of f_i will be of the form, for example,

$$\phi_A = \mathbf{A}_A \cdot \frac{\partial \ln T}{\partial \mathbf{q}} + \mathsf{B}_A : \frac{\partial \mathbf{u}}{\partial \mathbf{q}} + n\mathbf{\mathfrak{D}}_{AB} \cdot \mathbf{d}_A \tag{10-23}$$

where

$$\begin{aligned} \mathbf{A}_A &= A_A(v_0)\mathbf{v}_0 \\ \mathsf{B}_A &= B_A(v_0)\mathbf{v}_0{}^0\mathbf{v}_0 \\ \mathbf{\mathfrak{D}}_{AB} &= \mathfrak{D}_{AB}\mathbf{v}_0 \end{aligned} \tag{10-24}$$

The evaluation of these quantities can be carried out by expanding them in Sonnine polynomials; but, as the procedure is identical to that for a pure system, we will not bother to do that here.

The form of the transport equations for a mixture can also be determined from the above results and, since these do differ from the corresponding equations for a pure system, we now turn our attention to this problem. The species diffusion velocities are, for example,

$$\begin{aligned} \mathbf{v}_{DA} &= \frac{1}{n_A}\int d\mathbf{v}\, f_{LMA}\phi_A\mathbf{v}_0 \\ &= \frac{1}{3n_A}\int d\mathbf{v}\, f_{LMA}v_0{}^2\left\{n\mathfrak{D}_{AB}\,\mathbf{d}_A + A_A\frac{\partial \ln T}{\partial \mathbf{q}}\right\} \\ &= \frac{n^2}{n_A\rho} m_B D_{AB}\,\mathbf{d}_A - \frac{1}{\rho_A} D_A{}^T\frac{\partial \ln T}{\partial \mathbf{q}} \end{aligned} \tag{10-25}$$

where we have introduced the binary diffusion coefficient

$$D_{AB} = \frac{\rho}{3nm_B}\int d\mathbf{v}\, f_{LMA}\mathfrak{D}_{AB}v_0{}^2 \geq 0 \tag{10-26}$$

and the thermal diffusion coefficient

$$D_A{}^T = -\frac{m_A}{3}\int d\mathbf{v}\, f_{LMA}A_Av_0{}^2 \lessgtr 0 \tag{10-27}$$

In general, $\mathbf{v}_{DA} \neq \mathbf{v}_{DB}$, and there will be a net macroscopic flow of one of the species through the system; this is called diffusion. From the above results we see that there are several factors which will contribute to diffusion in a system. First, there will be a component of diffusion velocity in the $\mathbf{d}_A$ direction, which is a result of pressure and density gradients in the system. These effects were anticipated and observed long before the results of Chapman and Enskog were obtained. We also see that there will be a component of the diffusion velocity in either the $(\partial T/\partial \mathbf{q})$ or $-(\partial T/\partial \mathbf{q})$ direction. The possibility of this latter diffusion process, which is due to temperature gradients in the system, was not anticipated, and the successful prediction of this phenomenon, which was then subsequently observed experimentally, can be considered as one of the great triumphs of the Boltzmann equation theory. It is interesting to note that this phenomenon, which is referred to as thermal diffusion, does not occur in mixtures of Maxwell molecules, thus possibly preventing its earlier discovery by Maxwell who restricted his later work in kinetic theory to systems of these molecules.

It is clear from the form of the ϕ_i that the stress tensor will be of the usual form. This follows from simple parity considerations which indicate that only that component of ϕ_i containing $\mathbf{B}_i$ contributes to $\mathbf{P}$. A similar argument shows that the heat flux $\mathbf{Q}$ will contain a component in addition to the familiar conductive component arising from the temperature gradient. After a bit of algebraic manipulation, the heat flux can be written in the following form:

$$\mathbf{Q} = -\lambda \frac{\partial T}{\partial \mathbf{q}} + \frac{5}{2} kT \sum_{i=A,B} n_i \mathbf{v}_{Di} + \frac{kT}{n} \sum_{i,j=A,B} \frac{n_j D_i^T}{m_i D_{ij}} (\mathbf{v}_{Di} - \mathbf{v}_{Dj}) \quad \text{(10–28)}$$

with $D_{ii} \equiv 0$, and λ defined as

$$\lambda = -\sum_{i=A,B} \frac{R_i}{3} \int d\mathbf{v}\, f_{LMi} v_0^2 A_i \left(\frac{v_0^2}{2R_iT} - \frac{5}{2}\right) - \frac{k}{2n} \sum_{i,j=A,B} \frac{n_i n_j}{D_{ij}} \left[\frac{D_i^T}{\rho_i} - \frac{D_j^T}{\rho_j}\right] \quad \text{(10–29)}$$

We see that the three mechanisms which produce the heat flow are temperature gradients, species transport of energy relative to the average mass fluid flow, and transport due to thermal diffusion. The latter two are examples of reciprocal effects, that is, a cross-coupling of the fluxes and their driving forces, which occur in the system. These effects also occur in pure systems, but only in the third Chapman–Enskog approximation.

10–3 LINEARIZED AND MODEL BOLTZMANN EQUATIONS

As we saw in the last chapter, the linearized and model Boltzmann equations often offer the most practical basis for the consideration of flow problems. We can therefore anticipate that similar equations will play an equally important role in the kinetic theory of mixtures. There is considerably more latitude in the choice of a linearization procedure in the case of a mixture than for a pure system. In the latter case f_M and f_{LM} are usually the candidates for the unperturbed component of f. In a mixture, however, it is possible to linearize about a local Maxwellian which contains $\mathbf{u}$, T, or alternatively, we can introduce distinct species flow velocities and temperatures and linearize about local Maxwellians which contain these quantities. The theory of the collision operators which results in the latter case is complicated by the fact that these operators will not have the usual symmetry properties. This problem is not encountered if the first choice is elected. When one species is considerably heavier than the other, or present at considerably higher concentration, then self-collisions tend to play a dominant role, so that the species first equilibrate individually, and only then mutually. In this case the choice of separate local Maxwellians should be made.

In the spirit of the simple BGK model equation we can postulate the following model equations for a mixture:

$$\begin{aligned}\frac{\partial f_A}{\partial t} + \mathbf{v}\cdot\frac{\partial f_A}{\partial \mathbf{q}} &= \nu_A(f_{LMA} - f_A) + \nu_{AB}(f_{LMAB} - f_A)\\ \frac{\partial f_B}{\partial t} + \mathbf{v}\cdot\frac{\partial f_B}{\partial \mathbf{q}} &= \nu_B(f_{LMB} - f_B) + \nu_{BA}(f_{LMBA} - f_B)\end{aligned} \qquad (10\text{–}30)$$

In addition to the three adjustable parameters ν_A, ν_B, $\nu_{AB} = (n_B/n_A)\nu_{BA}$ which explicitly appear in the above equations, there are also the parameters T_{AB}, T_{BA}, $\mathbf{u}_{AB}$, $\mathbf{u}_{BA}$ which appear in f_{LMAB} and f_{LMBA}. The f_{LMi} are characterized by distinct n_i, T_i, $\mathbf{u}_i$, and the f_{LMij} are defined in terms of the four parameters listed above as

$$\begin{aligned}f_{LMAB} &= n_A\left(\frac{1}{2\pi R_A T_{AB}}\right)^{3/2} e^{-(\mathbf{v}-\mathbf{u}_{AB})^2/2R_AT_{AB}}\\ f_{LMBA} &= n_B\left(\frac{1}{2\pi R_B T_{BA}}\right)^{3/2} e^{-(\mathbf{v}-\mathbf{u}_{BA})^2/2R_BT_{BA}}\end{aligned} \qquad (10\text{–}31)$$

The above model was first postulated by Gross and Krook, and subsequently the linearized version was derived from the linearized Boltzmann equations by Sirovich using the Gross–Jackson technique described in Section 8–2. Although the constants T_{AB}, T_{BA}, $\mathbf{u}_{AB}$, $\mathbf{u}_{BA}$ can be chosen arbitrarily, they should be chosen in such a manner as to enhance the accuracy of the model description. A reasonable process for making such a choice has been suggested by Morse, who noted that the relaxation equations for momentum and temperature may be derived using both the full Boltzmann and model equations. In the first case, we find

$$\frac{\partial}{\partial t}(\mathbf{u}_B - \mathbf{u}_A) = -\alpha\left(\frac{1}{\rho_B} - \frac{1}{\rho_A}\right)\left(\frac{m_A + m_B}{2}\right)(\mathbf{u}_B - \mathbf{u}_A)$$

$$\frac{\partial}{\partial t}(T_B - T_A) = -\alpha\left[(T_B - T_A)\left(\frac{1}{n_B} + \frac{1}{n_A}\right) + \frac{(u_B - u_A)^2}{3k}\left(\frac{m_B}{n_A} - \frac{m_A}{n_B}\right)\right] \quad (10\text{–}32)$$

whereas the model equations give

$$\frac{\partial}{\partial t}(\mathbf{u}_B - \mathbf{u}_A) = \nu_{BA}(\mathbf{u}_{BA} - \mathbf{u}_B) - \nu_{AB}(\mathbf{u}_{AB} - \mathbf{u}_A)$$

$$\frac{\partial}{\partial t}(T_B - T_A) = n_A\nu_{AB}\left[\left(\frac{T_{BA}}{n_B} - \frac{T_{AB}}{n_A}\right) - \left(\frac{T_B}{n_B} - \frac{T_A}{n_A}\right) + \frac{m_A m_B}{3k(m_A + m_B)}(u_B - u_A)^2\left(\frac{m_A}{n_B} - \frac{m_B}{n_A}\right)\right] \quad (10\text{–}33)$$

For Maxwell molecules α is a constant, and we will consider this case here. By requiring the ratio of the momentum difference relaxation time to the temperature difference relaxation to be equal for both of the above equations we can determine the parameters $T_{AB}, \cdots, \mathbf{u}_{BA}$, and we find

$$\mathbf{u}_{AB} = \mathbf{u}_{BA} = \frac{m_A u_A + m_B u_B}{m_A + m_B}$$

$$T_{AB} = T_A + \frac{2m_A m_B}{(m_A + m_B)^2}\left[(T_B - T_A) + \frac{m_B}{6k}(u_B - u_A)^2\right]$$

$$T_{BA} = T_B + \frac{2m_A m_B}{(m_A + m_B)^2}\left[(T_A - T_B) + \frac{m_A}{6k}(u_B - u_A)^2\right] \quad (10\text{–}34)$$

10–4 THE BOLTZMANN FOKKER–PLANCK EQUATION

We will conclude the present chapter by considering an approximation of the Boltzmann equations for a mixture for a particular problem, the motion of a massive species at low concentration through a lighter, more copious species (called the solvent). This problem will serve to illustrate how the rigorous Boltzmann equation formalism can be used to substantiate less rigorously derived kinetic equations in the regime where both are applicable.

In 1828, in the course of reporting his studies on the motion of pollen particles in water, the English botanist Robert Brown commented on the irregular motions of the pollen particles which he had observed. As we now know, this Brownian motion results from the continuous interaction between the pollen (or Brownian) particles and the molecules of the solvent bath. Subsequent work on this problem by Einstein, Smoluchowski, and Langevin, to mention only a few of the well-known scientists whose interest has been stimulated, has led to a well developed probabilistic theory of Brownian motion. The probabilistic content of this theory is twofold, appearing both in the use of a distribution function $f_B(\mathbf{q},\mathbf{v},t)$ to describe the Brownian particles, and in the stochastic representation of the forces acting on the Brownian particles. The equation for $f_B(\mathbf{q},\mathbf{v},t)$ which results from this theory, called the Fokker–Planck equation, is

$$\frac{\partial f_B}{\partial t} + \mathbf{v} \cdot \frac{\partial f_B}{\partial \mathbf{q}} = \xi \left[\frac{\partial}{\partial \mathbf{v}} \cdot \left(\mathbf{v} f_B + R_B T \frac{\partial}{\partial \mathbf{v}} f_B \right) \right] \tag{10–35}$$

where ξ, the friction coefficient, depends phenomenologically on the solvent properties. The Fokker–Planck equation has a wide application in mathematics and physics; for example, it can be used to describe an electron gas, where a single electron receives many simultaneous but weak interactions in a typical microscopic time interval.

Clearly we can make use of the Boltzmann equations for a mixture to describe Brownian motion in the case where the solvent is a rarefied gas. In doing so we replace the stochastic description of the forces acting on a Brownian particle by a deterministic description according to Newton's laws of motion. This will furnish a test of the assumptions which go into obtaining the Fokker–Planck equation, and also provide an exact expression, in terms of molecular properties, for the friction coefficient. To formulate the Brownian motion problem in terms of the Boltzmann equations for a mixture, we consider a mixture of heavy Brownian (B) particles moving through a solvent of much lighter (A) particles; specifically, we consider $m_A/m_B \ll 1$. The concentration of the Brownian particles will be taken to be sufficiently low so that collisions between them can be neglected. Finally, the solvent particles will be assumed to be in equilibrium. The Boltzmann equations for this mixture then reduce to the single, linear equation

$$\begin{aligned} \frac{\partial f_B(\mathbf{v}_1)}{\partial t} + \mathbf{v}_1 \cdot \frac{\partial f_B(\mathbf{v}_1)}{\partial \mathbf{q}} &= \int d\mathbf{v}_2\, d\chi\, d\epsilon \sin\chi\, \sigma(\chi,V) V \{ f_B(\mathbf{v}_1') f_{MA}(v_2') \\ &\qquad - f_B(\mathbf{v}_1) f_{MA}(v_2) \} \\ &= J_{AB}(f_B) \end{aligned} \tag{10–36}$$

where we have used the differential cross-section notation introduced in Section 3–6 for reasons that will appear obvious later, and have suppressed the $\mathbf{q}$ and t dependence of f_B as a notational convenience. Writing $f_B = f_{MB}g$, we can then rewrite the collision term as

$$f_{MB}L_B(g) = f_{MB}\int d\mathbf{v}_2\, f_{MA}(\mathbf{v}_2)\int d\epsilon\, d\chi \sin\chi V\sigma(\mathbf{x},V)[g(\mathbf{v}_1') - g(\mathbf{v}_1)] \quad \textbf{(10–37)}$$

This term bears no resemblance at all to the Fokker–Planck collision term; however, as we shall soon see, that term is actually embedded in Equation (10–37). To see this we must exploit the fact that m_A/m_B is the natural small parameter appearing in $L_B(g)$, and expand in this parameter.

The quantities appearing in $L_B(\phi)$ which we will expand are $[g(\mathbf{v}_1') - g(\mathbf{v}_1)]$, $V\sigma$ and the $\mathbf{v}_1' - \mathbf{v}_1$ terms which result from the first expansion. The first term is expanded about $\mathbf{v}_1$, giving

$$g(\mathbf{v}_1') - g(\mathbf{v}_1) = (\mathbf{v}_1' - \mathbf{v}_1)\cdot\frac{\partial g(\mathbf{v}_1)}{\partial \mathbf{v}_1} + \tfrac{1}{2}(\mathbf{v}_1' - \mathbf{v}_1)(\mathbf{v}_1' - \mathbf{v}_1) : \frac{\partial^2 g(\mathbf{v}_1)}{\partial \mathbf{v}_1 \partial \mathbf{v}_1} + \cdots \quad \textbf{(10–38)}$$

The higher-order terms can be neglected here since

$$\mathbf{v}_1' - \mathbf{v}_1 = \frac{-m_A}{m_A + m_B}(\mathbf{V}' - \mathbf{V}) = O(m_A/m_B) \quad \textbf{(10–39)}$$

The relative velocity $\mathbf{V}$ can be expanded in terms of the dimensionless velocities $\mathbf{w}_1$ and $\mathbf{w}_2$, whose magnitudes will be of comparable order (assuming $\mathbf{w}_1$ is essentially given by its equipartition value), since

$$V = \left(\frac{2kT}{m_A}\right)^{1/2} w_2\left[1 - 2\left(\frac{m_A}{m_B}\right)^{1/2}\frac{w_1}{w_2}\cos\phi + \frac{m_A}{m_B}\frac{w_1^2}{w_2^2}\right]^{1/2} \quad \textbf{(10–40)}$$

where ϕ is the angle between $\mathbf{w}_1$ and $\mathbf{w}_2$. To order $(m_A/m_B)^{1/2}$ we then have

$$V\sigma(\chi, V) = \left(\frac{2kT}{m_A}\right)^{1/2} w_2\sigma\left(\chi, \left(\frac{2RT}{m_A}\right)^{1/2} w_2\right) \left[1 - \left(\frac{m_A}{m_B}\right)^{1/2}\frac{w_1}{w_2}\cos\phi\left(1 + \frac{w_2}{\sigma}\frac{\partial\sigma}{\partial w_2}\right)\right] \quad \textbf{(10–41)}$$

If we carry out the $\mathbf{v}_2$ integration in a polar coordinate system whose polar axis is along $\mathbf{v}_1$, we can then do the integration over the azimuthal angle ω by noting the following relationships:

$$\int_0^{2\pi} d\omega\,(\mathbf{w}_2 - \mathbf{w}_1) = 2\pi\left(\frac{w_2}{w_1}\cos\phi - 1\right)\mathbf{w}_1$$

$$\int_0^{2\pi} d\omega\,(\mathbf{w}_2 - \mathbf{w}_1)(\mathbf{w}_2 - \mathbf{w}_1) = 2\pi\mathbf{w}_1\mathbf{w}_2\left[\left(\frac{w_2}{w_1}\cos\phi - 1\right)^2 - \frac{1}{2}\frac{w_2^2}{w_1^2}\sin^2\phi\right] + \tfrac{1}{2}w_2^2\sin^2\phi\,\delta \quad (10\text{–}42)$$

We can also use similar relationships to carry out the ϵ integration, since ϵ is the azimuthal angle between $\mathbf{V}$ and $\mathbf{V}'$ (in this case the coefficients V'/V and $(V'/V)^2$ equal 1, and the polar angle is χ).

Substituting Equations (10–38), (10–39), and (10–41) into Equation (10–37), carrying out the ϵ, ω, and ϕ integrations, and returning to dimensional velocity variables, we finally obtain, after some lengthy manipulation,

$$J_{AB}(f_B) = \xi\frac{\partial}{\partial\mathbf{v}}\cdot\left(\mathbf{v}f_B + R_BT\frac{\partial}{\partial\mathbf{v}}f_B\right) + 0\left(\frac{m_A}{m_B}\right)^{3/2} \quad (10\text{–}43)$$

where the friction coefficient is given explicitly as

$$\xi = \tfrac{16}{3}\pi^{1/2}n_B\left(\frac{m_A}{m_B}\right)\left(\frac{1}{2R_AT}\right)^{5/2}\int_0^{\pi} d\chi\,\sin\chi(1 - \cos\chi)\int_0^{\infty} dv_2\,v_2^5e^{-v_2^2/2R_AT}\sigma(\chi,V) \quad (10\text{–}44)$$

We thus see that the Fokker–Planck equation can be viewed as a well defined approximation of the Boltzmann equation governing the Brownian motion problem. More important, we have explicitly determined the friction coefficient in terms of the molecular properties of the Brownian particle–solvent system.

References

The standard references for Sections 10–1 and 10–2 are:

1. S. Chapman and T. G. Cowling (1952). See General References.
2. J. O. Hirschfelder, C. F. Curtiss, and R. B. Bird, *Molecular Theory of Gases and Liquids*. New York: Wiley, 1954.

Model equations for mixtures were proposed by:

3. E. P. Gross and M. Krook, *Phys. Rev.*, vol. 102, p. 593, 1956.

4. B. Hamel, *Phys. Fluids*, vol. 8, p. 418, 1964.

5. L. W. Holway, Jr., *Phys. Fluids*, vol. 9, p. 1658, 1966.

6. T. Morse, *Phys. Fluids*, vol. 7, p. 2012, 1964.

7. L. Sirovich, *Phys. Fluids*, vol. 5, p. 917, 1962.

A good selection of papers on the theory of Brownian motion is contained in:

8. N. Wax, Ed., *Selected Papers on Noise and Stochastic Processes.* New York: Dover, 1954.

The Fokker–Planck form of the Boltzmann equation was first demonstrated by:

9. C. S. Wang Chang and G. E. Uhlenbeck, "The kinetic theory of a gas in alternating outside force fields: A generalization of the Rayleigh problem," University of Michigan, Ann Arbor, Mich., Engineering Research Institute Report, Project 2457, October 1956 (unpublished).

Problems

10–1. Write the full set of simultaneous Boltzmann equations for a three-component mixture.

10–2. Consider a space-uniform binary mixture, and define

$$H_A = \int d\mathbf{v}\, f_A \ln f_A$$

Prove that this function need not necessarily decay monotonely to equilibrium, and explain why it doesn't.

10–3. Demonstrate that an H theorem can be proved for the simultaneous Boltzmann equations for a binary mixture by using the H function indicated in the text.

10–4. Consider a chemically reacting binary mixture in which the components are monatomic and diatomic forms of the same atom. Write the simultaneous Boltzmann equations for the mixture in the case where C_A, the total increase per unit volume per unit time of atomic species, is constant.

10–5. Write the species and composite continuity equations for the reacting mixture considered in Problem 10–4. What are the summational invariants for this system?

10–6. Fick's second law of diffusion states that for a two-component system maintained at constant pressure and temperature in a stationary container, and with diffusion taking place in one direction only (say the x direction),

$$\frac{\partial n_A}{\partial t} = \frac{\partial}{\partial x} D_{AB} \frac{\partial n_A}{\partial x}$$

Prove that our results confirm this equation.

10–7. Determine the species momentum conservation equations for a binary mixture and show that when they are added together Equation (10–15) is obtained.

10–8. Show that the simple BGK equations for a mixture reproduce the conservation equations.

10–9. Determine a set of BGK-like equations for a three-component mixture.

10–10. Compute the friction coefficient when the Brownian and solvent fluid particles are hard spheres, not necessarily of the same size.

CHAPTER 11

generalized boltzmann equations

11-1 INTRODUCTION

There are many examples of phenomena which can be studied by means of an equation having the same structure as Boltzmann's equation, that is, an equation which equates the total time derivative of some distribution function to certain functionals of that distribution function which represent gain and loss terms resulting from the basic mechanism of the process being described. Some of these equations, which are used in describing gaseous systems not included in the BGL, are *ad hoc* modifications of Boltzmann's equation tailored to the problem at hand. In this chapter we will look at two equations of this type, the Wang Chang–Uhlenbeck equation, which is used to describe systems of structured particles, and the Enskog equation, which is used to describe dense gases. The use of Boltzmann-like equations is not confined to gas theory; in fact, it is not even restricted to the physical sciences. In concluding this chapter we briefly consider one of the more interesting applications of Boltzmann-like equations outside the traditional purview of the physical sciences. This is found in the theory of automobile traffic flow. The use of a Boltzmann equation is somewhat speculative here, since automobiles are guided by the whims of their drivers rather than the more predictable laws of nature, but then Boltzmann's equation itself was considered speculative at its inception.

11-2 THE WANG CHANG–UHLENBECK EQUATION

Our earlier discussion has been restricted to systems composed of monatomic molecules. When the constituent molecules are polyatoms, having a definite structure, we will have to consider nontranslational degrees of freedom, and consequently nontranslational, or internal, modes of energy. Generally, systems of polyatoms must be described quantum mechanically; however, even if a completely classical model is used it is clear that Boltzmann's equation cannot be used to furnish the kinetic description without the use of some major modifications.

Many of the important properties of systems of polyatoms can be studied by considering a semiclassical description in which the translational degrees of freedom are described classically and the internal degrees of freedom are described quantum mechanically. In this case we consider the distribution function $f_i(\mathbf{q},\mathbf{v},t)$, where the subscript i denotes the set of quantum numbers necessary to specify the internal state of the molecule. For rarefied gases it is not unreasonable to assume that f_i will obey the following Boltzmann-like equation:

$$\frac{\partial f_{1i}}{\partial t} + \mathbf{v}_1 \cdot \frac{\partial f_{1i}}{\partial \mathbf{q}} = J_{\mathrm{WCU}}(f) = \frac{1}{m} \sum_{j,k,l} \int d\mathbf{v}_2\, d\Omega\, [\sigma(\chi,V';kl,ij)V'f'_{1k}f'_{2l} - \sigma(\chi,V;ij,kl)Vf_{1i}f_{2j}] \quad \textbf{(11-1)}$$

Here $\sigma(\chi,V;ij,kl)$ is the cross section for scattering at the solid angle $d\Omega = \sin\chi\, d\chi\, d\epsilon$ (Figure 3–7) in which the pre-collisional internal states are i, j, and the post-collisional internal states are k, l. If we now introduce the assumption that

$$\sigma(\chi,V';kl,ij)V' = \sigma(\chi,V;ij,kl)V \quad \textbf{(11-2)}$$

then Equation (11–1) can be rewritten in the following simplified form:

$$\frac{\partial f_{1i}}{\partial t} + \mathbf{v}_1 \cdot \frac{\partial f_{1i}}{\partial \mathbf{q}} = \frac{1}{m} \sum_{j,k,l} \int d\mathbf{v}_2\, d\Omega\, \sigma(\chi,V;ij,kl)V[f'_{1k}f'_{2l} - f_{1i}f_{2j}] \quad \textbf{(11-3)}$$

The above equation was first postulated by Wang Chang and Uhlenbeck and is referred to as the Wang Chang–Uhlenbeck (WCU) equation (a similar equation was developed independently by deBoer). Since we have not rigorously derived the WCU equation, it is not immediately apparent what restrictions must be placed on its use. It can be shown that this equation gives a very good description of polyatomic systems in which the interparticle potential is spherically symmetric. Thus we must exclude polyatoms possessing significant dipole moments from our considerations. The WCU equation also proves inadequate in describing phenomena associated with the presence of magnetic fields, where the transport coefficients take on a tensorial form. A further shortcoming of the WCU equation is that the scattering cross section σ can only be determined by solving a quantum mechanical scattering problem. In general, these quantities are not known, and so the theory we are about to consider is therefore still incomplete.

As usual, the moments of f_i can be related to the thermo-fluid variables of interest. The density, fluid flow velocity, and pressure tensor are defined essentially as before, the difference being that we must now also sum over internal states in addition to integrating when forming moments.

$$\sum_i \int d\mathbf{v}\, f_i \begin{pmatrix} 1 \\ \mathbf{v} \\ \mathbf{v}_0\mathbf{v}_0 \end{pmatrix} = \begin{pmatrix} \rho \\ \rho\mathbf{u} \\ \mathsf{P} \end{pmatrix} \tag{11–4}$$

The fluid internal energy density per unit mass, ϵ, is now defined as

$$\begin{aligned} \rho\epsilon &= \sum_i \int d\mathbf{v}\, f_i(\tfrac{1}{2}(\mathbf{v}-\mathbf{u})^2 + \epsilon_i) \\ &\equiv \rho(\epsilon_T + \epsilon_I) \end{aligned} \tag{11–5}$$

where ϵ_i is the energy per unit mass of the ith quantum state, and ϵ_T, ϵ_I are the translational and internal energy densities.

Moment equations can be formed by multiplying the WCU equation by 1, $\mathbf{v}$, and $(\frac{1}{2}v_0^2 + \epsilon_i)$ and operating with $\sum_i \int d\mathbf{v}$. The resulting equations are

$$\begin{gathered} \frac{\partial \rho}{\partial t} + \frac{\partial}{\partial \mathbf{q}} \cdot \rho\mathbf{u} = 0 \\ \rho\left(\frac{\partial \mathbf{u}}{\partial t} + \mathbf{u}\cdot\frac{\partial \mathbf{u}}{\partial \mathbf{q}}\right) + \frac{\partial}{\partial \mathbf{q}} \cdot \mathsf{P} = 0 \\ \frac{\partial}{\partial t}\rho(\epsilon + \tfrac{1}{2}u^2) + \frac{\partial}{\partial \mathbf{q}} \cdot (\rho\mathbf{u}(\epsilon + \tfrac{1}{2}u^2) + \mathbf{u}\cdot\mathsf{P} + \mathbf{Q}_T + \mathbf{Q}_I) = 0 \end{gathered} \tag{11–6}$$

where the heat flux has been separated into a translational energy and an internal energy flow component according to

$$\begin{pmatrix} \mathbf{Q}_T \\ \mathbf{Q}_I \end{pmatrix} = \sum_i \int d\mathbf{v}\, f_i \mathbf{v}_0 \begin{pmatrix} v_0^2/2 \\ \epsilon_i \end{pmatrix} \tag{11–7}$$

In order to demonstrate the approach to equilibrium and determine the equilibrium state we again consider the proof of an H theorem. The H function is now defined as

$$H = \sum_i \int d\mathbf{v}\, f_i \ln f_i \tag{11–8}$$

and it is an easy matter to show, through the usual manipulations, that H decays monatonely to its equilibrium value. The form of the equilibrium distribution function is

$$f_{Mi} = \frac{\rho}{(2\pi RT)^{3/2}} \frac{e^{-1/RT(v_0^2/2+\epsilon_i)}}{Z} \tag{11–9}$$

where Z is the partition function of equilibrium statistical mechanics, defined here as

$$Z = \sum_j e^{-\epsilon_j/RT} \tag{11–10}$$

The temperature T which appears in the above equations is defined, with the help of the constant volume heat capacities for translational and internal degrees of freedom, as

$$\epsilon = (C_v{}^T + C_v{}^I)T = C_vT \tag{11–11}$$

11–3 NORMAL SOLUTIONS OF THE WCU EQUATION

The form of the transport equations can be uncovered by considering the normal solutions of the WCU equation. In constructing these solutions we will want to distinguish between the two distinct cases of facile and nonfacile transfer of energy between translational and internal modes. We shall only consider the first case here, reserving the nonfacile case for the problems at the end of the chapter.

When the collision frequencies of elastic and inelastic collisions are of the same order of magnitude, then there will be a facile transfer of energy between the translational and internal degrees of freedom. Accordingly, it is reasonable to describe the system, in the lowest order approximation, with just one temperature, and we can impose the following constraints on the $f_i{}^{(j)}$:

$$\sum_i \int d\mathbf{v}\, f_i{}^{(j)} \begin{pmatrix} 1 \\ \mathbf{v} \\ (\frac{1}{2}v_0{}^2 + \epsilon_i) \end{pmatrix} = \begin{pmatrix} \rho \\ \rho\mathbf{u} \\ \rho(C_v{}^T + C_v{}^I)T \end{pmatrix} \delta_{0j} \tag{11–12}$$

The procedure for determining the $f_i^{(j)}$ is by now familiar, and we will only indicate the pertinent results here. The first approximation is the local Maxwellian

$$f_i^{(0)} = f_{LMi} = \frac{\rho}{(2\pi RT)^{3/2}Z} e^{-1/RT(v_0^2/2+\epsilon_i)} \tag{11-13}$$

In the next approximation setting $f_i^{(1)} = f_{LMi}\phi_i$, we find the following equation for ϕ_i:

$$\begin{aligned}\rho^2 I_{WCU}(\phi_i) = f_{LMi}\Bigg[(w_0^2 - \tfrac{5}{2})\mathbf{v}_0 \cdot \frac{\partial \ln T}{\partial \mathbf{q}} &+ 2\left(w_{0\alpha}w_{0\beta} - \frac{w_0^2}{3}\delta_{\alpha\beta}\right)\frac{\partial u_\alpha}{\partial q_\beta} + \left(\frac{\epsilon_i}{RT} - \frac{\epsilon_I}{RT}\right)\mathbf{v}_0 \cdot \frac{\partial \ln T}{\partial \mathbf{q}} \\ &+ \frac{C_v^I}{C_v}\left(\frac{v_0^2}{3RT} - \frac{\epsilon_i}{C_v^I T} + \frac{\epsilon_I}{C_v^I T} - 1\right)\frac{\partial}{\partial \mathbf{q}} \cdot \mathbf{u}\Bigg]\end{aligned} \tag{11-14}$$

The linear isotropic operator I_{WCU} is similar to the I operator used in Chapter 6 [see Equation (6–44)]; the inhomogeneous term which appears in the above integral equation contains all the terms that appear in the corresponding equation for a monatomic gas plus two additional terms which correspond to the transport and exchange of internal energy. From the above equation we can infer that ϕ_i will be of the form [see Equation (6–55)]:

$$\begin{aligned}\phi_i = \frac{1}{\rho}(2RT)^{1/2}\mathbf{A}_T \cdot \frac{\partial \log T}{\partial \mathbf{q}} &+ \frac{1}{\rho}(2RT)^{1/2}\mathbf{A}_I \cdot \frac{\partial \log T}{\partial \mathbf{q}} \\ &+ \frac{1}{\rho}\mathsf{B} : \frac{\partial \mathbf{u}}{\partial \mathbf{q}} + \frac{1}{\rho}\frac{C_v^I}{C_v} C \frac{\partial}{\partial \mathbf{q}} \cdot \mathbf{u}\end{aligned} \tag{11-15}$$

with

$$\begin{aligned}\mathbf{A}_T &= A_T(w_0,\epsilon_i)\mathbf{w}_0 \\ \mathbf{A}_I &= A_I(w_0,\epsilon_i)\mathbf{w}_0 \\ \mathsf{B} &= B(w_0,\epsilon_i)\mathbf{w}_0{}^0\mathbf{w}_0 \\ C &= C(w_0,\epsilon_i)\end{aligned} \tag{11-16}$$

The above relationships are a direct consequence of the linear isotropic form of I_{WCU}. If I_{WCU} were not isotropic we could find, for example, that in the presence of a magnetic field the A's, and consequently the thermal conductivity, are tensor quantities. The isotropy of the collision operator is a direct consequence of the use of collision cross sections rather than scattering amplitudes in the Boltzmann-like equation (11–1), and is the reason why the WCU equation does not correctly describe nonscalar transport phenomena.

The heat flux vectors and the stress tensor can be calculated from Equations (11–15) and (11–16), and we find

$$\mathbf{Q}_T = \frac{-2R^2T}{3}[\mathbf{A}_T,(\mathbf{A}_T + \mathbf{A}_I)]_{\text{WCU}} \frac{\partial T}{\partial \mathbf{q}}$$

$$\equiv -\lambda_T \frac{\partial T}{\partial \mathbf{q}} \qquad (11\text{–}17)$$

$$\mathbf{Q}_I = \frac{-2R^2T}{3}[\mathbf{A}_I,(\mathbf{A}_T + \mathbf{A}_I)]_{\text{WCU}} \frac{\partial T}{\partial \mathbf{q}}$$

$$\equiv -\lambda_I \frac{\partial T}{\partial \mathbf{q}} \qquad (11\text{–}18)$$

$$\mathsf{P} = \frac{-RT}{10}[\mathsf{B},\mathsf{B}]_{\text{WCU}} 2\left(\mathsf{D} - \frac{D_{\alpha\alpha}}{3}\mathsf{U}\right) - \left(\frac{C_v^I}{C_v}\right)^2 RT[C,C]_{\text{WCU}}\mathsf{U}\frac{\partial}{\partial \mathbf{q}}\cdot\mathbf{u}$$

$$\equiv -2\mu\left(\mathsf{D} - \frac{D_{\alpha\alpha}}{3}\mathsf{U}\right) - \mu_B \frac{\partial}{\partial \mathbf{q}}\cdot\mathbf{u}\mathsf{U} \qquad (11\text{–}19)$$

where the bracket operators are similar to those introduced earlier for monatomic gases [see Equation (6–59)]. Specifically, if X and Y are tensors of the same order,

$$[\mathsf{X},\mathsf{Y}]_{\text{WCU}} = -\sum_i \int d\mathbf{v}\, \mathsf{X} \odot I_{\text{WCU}}(\mathsf{Y})$$

$$= -\sum_i \int d\mathbf{v}\, \mathsf{Y} \odot I_{\text{WCU}}(\mathsf{X}) \qquad (11\text{–}20)$$

where $\odot$ indicates a scalar product. The rate of strain tensor which appears in Equation (11–19) was defined earlier [see Equation (6–61)]. We see that in the present approximation the heat flux is given by Fourier's law with $\lambda = \lambda_T + \lambda_I$, but the stress tensor now includes a contribution from bulk viscosity effects. It can be shown that μ_B is related to the relaxation time which characterizes the equilibration of the translational and internal degrees of freedom.

11–4 MODEL EQUATIONS FOR STRUCTURED PARTICLES

In light of the problems encountered in using the WCU equation to generate numerical results it is not surprising that model equations play an especially important role in the theory of polyatoms. Modeling can be employed at two levels. We can consider a completely classical description in which the molecules themselves are modeled; for example, assume that they are rigid rotating ovaloids, in which case rotational as well as translational motions will have to be considered. There is also the possibility of modeling within the framework of the semiclassical description, and replacing the WCU equation with a BGK-type equation.

Many models of the first type mentioned above have been considered. The simplest of these is the so-called perfect rough sphere, which was first postulated by G. Bryan in 1894. A considerable number of numerical results have been obtained for this model, and these are found to be in reasonable agreement with experimental results. The model molecules are rotating hard spheres which can transfer translational and rotational energy upon colliding. A collision is assumed to reverse the relative velocities of the points of contact; thus the relationship

$$\begin{aligned}\bar{\mathbf{V}} &= \bar{\mathbf{v}}_2 - r_0\boldsymbol{\alpha} \wedge \boldsymbol{\omega}_2 - \bar{\mathbf{v}}_1 - r_0\boldsymbol{\alpha} \wedge \boldsymbol{\omega}_1 \\ &= -\mathbf{v}_2 + r_0\boldsymbol{\alpha} \wedge \boldsymbol{\omega}_2 + \mathbf{v}_1 + r_0\boldsymbol{\alpha} \wedge \boldsymbol{\omega}_1 \qquad \textbf{(11–21)}\end{aligned}$$

together with the equations for the conservation of energy and linear and angular momentum serve to specify the pre-collisional velocities, $\bar{\mathbf{v}}_i$, and the pre-collisional angular velocities, $\bar{\boldsymbol{\omega}}_i$, in terms of the post-collisional values, $\mathbf{v}_i$, $\boldsymbol{\omega}_i$, and the configurational vector, $\boldsymbol{\alpha}$. One of the simplifying features of the model is that orientational coordinates do not have to be used to describe the molecular states. A feature of the model which seems to be peculiar to real polyatoms is that inverse collisions do not exist, and therefore the $\bar{\mathbf{v}}_i$, $\bar{\boldsymbol{\omega}}_i$ cannot be replaced by $\mathbf{v}_i'$, $\boldsymbol{\omega}_i'$ (the WCU equation presumes the existence of inverse collisions). Other classical molecular models which have been utilized to represent structured particles are loaded rigid spheres, which are hard spheres whose mass center is displaced from the geometrical center of the sphere, and sphero-cylinders, which are cylinders capped with hemispheres.

A BGK-type model representation of the WCU is especially appealing since adjustable parameters, which can be determined by comparison with experimental results, are used in place of expressions involving the unknown cross sections. An analog of the simple BGK equation has been obtained by Morse from the WCU equation through the use of the following arguments. First the WCU collision term is rewritten in the following form:

$$\begin{aligned}J_{\mathrm{WCU}}(f) &= \frac{1}{m}\sum_{j}\int d\mathbf{v}_2\, d\Omega\, \sigma(\chi,V;ij,ij)V\{f_{1i}'f_{2j}' - f_{1i}f_{2j}\} \\ &\quad + \frac{1}{m}\sum_{j,k,l}\int d\mathbf{v}_2\, d\Omega\, \sigma(\chi,V;ij,kl)V(1 - \delta_{ik}\delta_{jl}) \cdot \{f_{1k}'f_{2l}' - f_{1i}f_{2j}\} \\ &= J_{\mathrm{WCU}}{}^{E}(f) + J_{\mathrm{WCU}}{}^{I}(f) \qquad \textbf{(11–22)}\end{aligned}$$

The first term in the above expression is the collision term for elastic collisions, while the second term is the collision term for inelastic collisions. The elastic collision term can be written in the BGK form by making the usual arguments, so that we may make the replacement

$$J_{\mathrm{WCU}}{}^{E}(f) \to \nu_E(f_{\mathrm{LM}T_i} - f_i) \tag{11–23}$$

with

$$\nu_E = \frac{1}{m} \sum_j \int d\mathbf{v}_2\, d\Omega\, \sigma(\chi, V; ij, ij) V f_{2j} \tag{11–24}$$

$$f_{\mathrm{LM}T_i} = \rho \left(\frac{1}{2\pi R T_T}\right)^{3/2} e^{-v_0^2/2RT_T} \frac{e^{-\epsilon_i/RT_I}}{\sum_j e^{-\epsilon_j/RT_I}} \tag{11–25}$$

Separate temperatures have been used to characterize the internal and translational degrees of freedom in order to allow for the possibility that the transfer of energy between the internal and translational modes is nonfacile. In general, we will define

$$\begin{aligned} 3\rho R T_T &= \sum_i \int d\mathbf{v}\, f_i v_0^2 \\ \rho C_v{}^I T_I &= \sum_i \int d\mathbf{v}\, f_i \epsilon_i \end{aligned} \tag{11–26}$$

in terms of which T, the total system temperature, is defined as

$$T = \frac{(\frac{3}{2} R T_T + C_v{}^I T_I)}{\frac{3}{2} R + C_v{}^I} \tag{11–27}$$

A similar argument leads to the replacement of $J_{\mathrm{WCU}}{}^I(f_i)$ according to

$$J_{\mathrm{WCU}}{}^{I}(f) \to \nu_I(f_{\mathrm{LM}Ii} - f_i) \tag{11–28}$$

with

$$\nu_I = \sum_{j,k,l} (1 - \delta_{ik}\delta_{jl}) \int d\mathbf{v}_2\, d\Omega\, \sigma(\chi, V; ij, kl) V f_{2j} \tag{11–29}$$

$$f_{\mathrm{LM}I_i} = \rho \frac{1}{(2\pi RT)^{3/2}} e^{-v_0^2/2RT} \frac{e^{-\epsilon_i/kT}}{Z} \tag{11–30}$$

The particular choice of local Maxwellians is made to insure that the model collision term gives the proper moment equations. A Gross–Jackson-type scheme can also be used to generate generalized model equations for the WCU equation. The simple model equation has been used to treat a wide variety of problems involving system of polyatoms.

11-5 ENSKOG'S EQUATION FOR DENSE GASES

In Chapter 6 we were able to prove that the thermal conductivity and viscosity for a gas in the BGL are independent of density. Although the BGL is an idealization, it is reasonable to expect that it gives an accurate representation of rarefied gases. For denser gases, however, the BGL cannot be expected to hold, and in this regime the Boltzmann equation is no longer valid. The problem of generalizing Boltzmann's equation to include dense systems has received considerable attention in the past twenty years, but despite considerable efforts no general solution is yet in sight. For this reason the Boltzmann-like equation postulated by Enskog for dense systems of hard sphere molecules is particularly appealing, since it gives results that are in reasonable agreement with experiment.

Enskog's equation is a modified version of the Boltzmann equation in which certain obvious deficiencies of the latter in the dense gas regime are corrected for. First, the notion of localized collisions is abandoned, and the full spatial dependence of the distribution functions in the collision term is retained. This will allow us to describe the transport at a distance of momentum and energy, that is, transport of momentum and energy which is effected through the mechanism of the interparticle potential. In the BGL we only take into account kinetic transport, that is, momentum and energy transport associated with center of mass motion. However, in a dense gas this is no longer sufficient (see Figure 11–1) since both energy and momentum can be transported across an arbitrary plane by a given molecule without that molecule's center of mass crossing the plane. Enskog also abandoned the simple *stosszahlansatz* for about-to-collide particles, in which it is assumed that there are no correlations between the particles, by an *ansatz* which assumes that there is a definite correlation, and that this correlation is of the same functional form as is found in equilibrium. In equilibrium, we know from the methods of equilibrium statistical mechanics that for two particles at positions $\mathbf{q}$, $\mathbf{q} + \mathbf{r}$,

$$f_2(\mathbf{q},\mathbf{v}_1,\mathbf{q} + \mathbf{r},\mathbf{v}_2) = g(r)f(\mathbf{v}_1)f(\mathbf{v}_2) \tag{11–31}$$

that is, the entire correlation is through the function $g(r)$, the radial distribution function. This quantity depends only on the particle separation, and is a function of density. Enskog's *ansatz* replaces f_2 for non-equilibrium pre-collision states according to

$$f_2(\mathbf{q}_1, \mathbf{v}_1, \mathbf{q}_1 - 2\alpha r_0, \mathbf{v}_2, t) = g^*(\mathbf{q}_1 - \alpha r_0) \\ f(\mathbf{q}_1,\mathbf{v}_1,t)f(\mathbf{q}_1 - 2\alpha r_0, \mathbf{v}_2, t) \tag{11–32}$$

The quantity $g^*(\mathbf{q}_1 - \alpha r_0)$ is the hard sphere radial distribution function at contact, in which ρ is replaced by its local value at the point in space

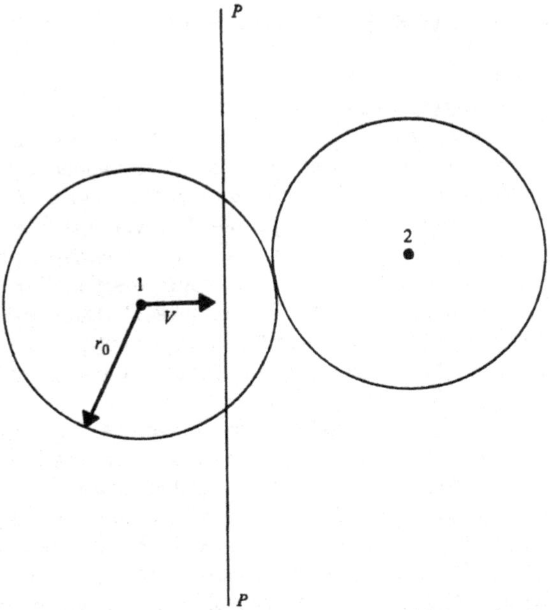

Figure 11–1 Particle 1 collides with particle 2. Although the center of 1 does not cross PP, momentum and energy are nevertheless transported across this plane. In the BGL $r_0 \to 0$, and this effect vanishes.

specified in the argument. To first order in the density we have, from equilibrium statistical mechanics,

$$g^*(\mathbf{r}) = 1 + \tfrac{5}{8}b\rho(\mathbf{r}) \tag{11–33}$$

where

$$b\rho(\mathbf{r}) = \tfrac{2}{3}\pi\rho(\mathbf{r})(2r_0)^3$$

Note that in the BGL, $g^* = 1$.

The Enskog equation, which incorporates the modifications discussed above into Boltzmann's equation, is then

$$\frac{Df_1}{Dt} = J_E(f) \equiv \frac{1}{m}\int d\mathbf{v}_2\, d\epsilon\, d\theta\, B(\theta,V)[g^*(\mathbf{q}_1 + \boldsymbol{\alpha} r_0)f(\mathbf{q}_1,\mathbf{v}_1{}',t)f(\mathbf{q}_1 + 2\boldsymbol{\alpha} r_0,\ \mathbf{v}_2{}',t)$$
$$- g^*(\mathbf{q}_1 - r_0\boldsymbol{\alpha})f(\mathbf{q}_1,\mathbf{v}_1,t)f(\mathbf{q}_1 - 2r_0\boldsymbol{\alpha},\ \mathbf{v}_2,\ t) \tag{11–34}$$

In solving the above equation by the Chapman–Enskog technique, only terms linear in $(\partial f/\partial \mathbf{q})$ are retained on the left-hand side in the first two approximations. In these approximations we can therefore expand the spatially dependent terms in J_E about $\mathbf{q}_1$ and retain only the first two terms without introducing any additional error. The Enskog collision term then becomes

$$J_E \to g^*(\mathbf{q}_1)J(f) + \frac{g^*(\mathbf{q}_1)}{m} \int d\mathbf{v}_2\, d\epsilon\, d\theta\, B(\theta,V) 2r_0\alpha \left\{ f_1' \frac{\partial f_2'}{\partial \mathbf{q}_1} + f_1 \frac{\partial f_2}{\partial \mathbf{q}_1} \right\}$$
$$+ \frac{1}{m} \int d\mathbf{v}_2\, d\epsilon\, d\theta\, B(\theta,V) r_0\alpha \cdot \frac{\partial g^*}{\partial \mathbf{q}_1}(\mathbf{q}_1)(f_1'f_2' + f_1f_2) \quad \textbf{(11-35)}$$

and it is this expression which is usually used in the further development of the Enskog equation.

11–6 A BOLTZMANN EQUATION FOR TRAFFIC FLOW

Boltzmann-like equations have found application in many areas of the traditional sciences. In our concluding section we consider a somewhat speculative application of a Boltzmann-like equation to a very contemporary problem, the flow of automobiles on a highway. This theory, which is due to Prigogine and co-workers, gives qualitative results in good agreement with "experiment." The Boltzmann equation which is used describes the behavior of the distribution function $f(x,v,t)$, where $f(x,v,t)\,dx\,dv$ is the number of autos at x, $x + dx$ having speed in the range v, $v + dv$ at time t (the traffic problem is essentially one-dimensional—until the helicopter age descends upon us in full force). It is postulated that f satisfies the following equation:

$$\frac{\partial f}{\partial t} + v\frac{\partial f}{\partial x} = \nu(f^0 - f) + \rho(u - v)(1 - P)f \qquad \textbf{(11-36)}$$

Here ν^{-1} is a characteristic relaxation time; f^0 is the desired distribution function;

$$\rho = \int_0^\infty dv\, f = \int_0^\infty dv\, f^0$$

is the vehicle density;

$$u = \rho^{-1} \int_0^\infty dv\, vf$$

is the average speed; and P is the probability of an auto passing another auto. The collision term consists of two distinct contributions. The first term is the limiting form of the collision term for zero density, and represents an exponential decay characterized by ν^{-1}, to the desired value of f. The second term attempts to take into account the presence of other autos on the road, which will tend to reduce the average traffic velocity.

As an example of how the techniques developed for treating Boltzmann's equation can also be applied to Boltzmann-like equations we will consider a ramification of applying the Chapman–Enskog method of solution to Equation (11–36). First, we note that there is only one summational invariant associated with the problem under consideration, namely the number of automobiles. The vehicle continuity equation is therefore of the usual form,

$$\frac{\partial \rho}{\partial t} + \frac{\partial}{\partial x} \rho u = 0 \tag{11–37}$$

In the Chapman–Enskog theory, $f(x,v,t)$ becomes replaced by $f(x,v;\rho(x,t))$, and accordingly, $u(x,t)$ becomes replaced by $u(x;\rho(x,t))$. We can thus write the vehicle continuity equation in the following form:

$$\frac{\partial \rho}{\partial t} + \frac{\partial}{\partial \rho} (\rho u) \frac{\partial \rho}{\partial x} = 0 \tag{11–38}$$

which indicates that

$$\rho = \rho\left(x - \frac{\partial \rho u}{\partial \rho} t\right) \tag{11–39}$$

corresponding to a wave propagating with velocity $u + (\partial \rho u/\partial \rho)$. This result supports the wave-like descriptions which have provided one of the main methods of studying traffic flow problems.

References

The semiclassical theory of structured particles was proposed by:

1. C. S. Wang Chang and G. E. Uhlenbeck, University of Michigan, Ann Arbor, Mich., Engineering Research Institute Report CM-681, 1951 (unpublished).

A published summary of this paper and of the unpublished work of J. deBoer are given in:

2. C. S. Wang Chang, G. E. Uhlenbeck, and J. deBoer, in *Studies in Statistical Mechanics*, vol. 2, J. deBoer and G. E. Uhlenbeck, Eds. New York: Wiley, 1964.

The WCU equation has been discussed by:

3. J. O. Hirschfelder, C. F. Curtiss, and R. B. Bird, *Molecular Theory of Gases and Liquid*. New York: Wiley, 1954.

4. L. Finkelstein and S. Harris, *Phys. Fluids*, vol. 9, p. 8, 1966.

5. L. Waldmann, in *Fundamental Problems in Statistical Mechanics II*, E. G. D. Cohen, Ed. New York: Wiley, 1968.

Classical models of structured particles are discussed in Reference 3 and in:

6. S. Chapman and T. G. Cowling (1952). See General References.

7. C. F. Curtiss and Ch. Muckenfuss, *J. Chem. Phys.*, vol. 26, p. 1619, 1957.

8. J. S. Dahler, *J. Chem. Phys.*, vol. 30, p. 1447, 1959.

9. N. Taxman, *Phys. Rev.*, vol. 110, p. 1235, 1958.

The BGK model for structured particles was proposed by:

10. T. Morse, *Phys. Fluids*, vol. 7, p. 159, 1964.

Generalized BGK models have been proposed by:

11. L. W. Holway, Jr., *Phys. Fluids*, vol. 9, p. 1658, 1966.

12. F. B. Hansen and T. Morse, *Phys. Fluids*, vol. 10, p. 345, 1967.

The Enskog equation for dense gases is discussed in References 3 and 6. It was originally proposed in:

13. D. Enskog, *Svensk. Akad. Hadl.*, vol. 63, 1922.

Generalizations of Boltzmann's equation for dense gases have also been considered in References 1 through 7 of Chapter 2.

The application of a Boltzmann-like equation to the theory of traffic flow is considered in

14. I. Prigogine, R. Herman, and R. Anderson, in *Vehicular Traffic Science*, L. Edie, R. Herman, and R. Rothery, Eds., New York, American Elsevier, 1967.

Problems

11–1. Determine the continuity equation satisfied by

$$\rho_i = \int d\mathbf{v}\, f_i$$

Why isn't this equation of the same form as the first of Equations (11–6)?

11–2. Prove the H theorem for the WCU equation.

11–3. Translational and internal temperatures are defined in Equation (11–26). Show that the first-order Chapman–Enskog solution of the WCU equation is characterized by equal translational and internal temperatures, but that this equality does not hold in the second Chapman–Enskog approximation.

11–4. In the second-order Chapman–Enskog solution to the WCU equation we have $T_T \neq T_I$. Show that the difference between these quantities is proportional to the coefficient of bulk viscosity, μ_B.

11–5. In the second-order Chapman–Enskog solution of the WCU equation, f_i can be written in terms of the quantities A_T, A_I, B, and C, which are to be determined by some approximate method. Write the equations which determine these quantities, and explain why it no longer suffices to expand solely in Sonnine polynomials to effect an approximate solution. How could these quantities be expanded?

11–6. When the exchange of energy between translational and internal modes is no longer facile, we write, as in Equation (11–22), $J_{\mathrm{WCU}} = J_{\mathrm{WCU}}{}^E + J_{\mathrm{WCU}}{}^I$, and order the terms according to $J_{\mathrm{WCU}}{}^E = O(\delta^{-1})$, $J_{\mathrm{WCU}}{}^I = O(\delta)$. Determine the first-order Chapman–Enskog solution in this case if the *ansatz* is made that in lowest order the internal energies are distributed according to

$$\frac{e^{-\epsilon_i/kT_I}}{\sum_i e^{-\epsilon_i/kT_I}}$$

11–7. Show that the simple BGK model of the WCU equation given in Equations (11–22) through (11–30) satisfies the conservation equations, and that an H theorem can be proved for it.

11–8. Determine the transformations that take $\tilde{\mathbf{v}}_1$, $\tilde{\mathbf{v}}_2$, $\tilde{\boldsymbol{\omega}}_1$, $\tilde{\boldsymbol{\omega}}_2$ into $\mathbf{v}_1$, $\mathbf{v}_2$, $\boldsymbol{\omega}_1$, $\boldsymbol{\omega}_2$ for perfect rough sphere molecules, and demonstrate that inverse collisions do not exist for this model.

11–9. Are the conservation equations for mass and momentum obtained from the Enskog equation the same that are obtained from the Boltzmann equation? Explain your answer.

11–10. The Chapman–Enskog procedure may be applied to the Enskog equation. The expressions in the collision term are ordered according to $g^*J(f) = O(\delta^{-1})$, and the other two terms are $O(\delta^0)$. What is the first-order solution?

11–11. In automobile traffic flow theory it is reasonable to assume that the probability of passing, P, is

$$P = \left(1 - \frac{\rho}{\rho^*}\right) \qquad \rho < \rho^*$$
$$P = 0 \qquad \rho \geq \rho^*$$

where ρ^* is the density at traffic jam conditions. Using this assumption, determine the stationary, uniform solution of Equation (11–36) for given f_0.

CHAPTER 12

comparison of theory with experiment

12–1 TRANSPORT COEFFICIENTS

We mentioned at the beginning of this book that there exists an important test for the theory that we subsequently developed, namely, the comparison of our theoretical results with corresponding experimentally derived results. Several types of experiments which probe the theory to various depths are possible. Perhaps the most superficial such test, but still a very important one, is that which results from comparing the theoretical and experimental values of the transport coefficients. As we have shown [see Equation (6–7)], the expressions we derived for the transport coefficients involve complicated averages of the distribution function, and it is only these averages that we test and not the distribution function itself. A particular advantage of considering such an average is that the results can be expected to be relatively insensitive to the details of the interparticle potential, and thus the uncertainty with which this quantity is known does not always lead to a concomitant numerical uncertainty. Thus, as we have noted [see Equation (6–82)], the relationship $\mu = (4/15R)\kappa$ holds to a very high degree of accuracy for monatomic gases. This result is independent of the assumed molecular model (and the temperature as well) and thus provides a minimal test to which the theory can be put. It is found that experimental results do indeed satisfy this relationship. Another test which does not depend on the choice of molecular model either is the determination of the density dependence of μ and κ. Our theoretical results predict that these quantities will be independent of density; this result was first verified experimentally by Maxwell. The regime in which the transport coefficients become density dependent is associated with dense gas effects, which as we saw in Equation (11–3) are not adequately described by Boltzmann's equation.

The temperature dependence of the transport coefficients is sensitive to the choice of molecular model. In Section 6–7 we found that for power law potentials [$\phi(R) \propto 1/R^{s-1}$], μ and κ are proportional to $T^{(1/2)+[2/(s-1)]}$. For hard spheres μ and κ would then be proportional to $T^{1/2}$; however, their actual temperature dependence as determined experimentally is somewhat stronger. This can be corrected for by modifying the hard sphere diameter so that it becomes a function of temperature, the justification for this being that at higher temperatures colliding particles will be, on the average, more energetic and thus penetrate further into each other's repulsive force field than at lower temperatures. A similarly motivated modification can be made for the general power law potential by taking s to be a function of temperature. However, in order to obtain truly satisfactory agreement with experiment over a wide range of conditions, it is necessary to take into account the attractive forces which act between the molecules at large separation. These are mainly induced dipole forces (often referred to as Van der Waals forces). These forces can be taken into account by representing $\phi(R)$ as a superposition of two separate power law potentials—a strong, short ranged repulsive potential and a weak, long ranged attractive potential:

$$\phi(R) = \frac{K}{R^{s-1}} - \frac{K'}{R^{s'-1}} \tag{12–1}$$

Interparticle potentials of the above form are referred to as Lennard–Jones potentials. Two versions of this model are commonly used. The first version, called the Sutherland model, treats the repulsive potential as a hard sphere potential, and the attractive potential as a power law potential. The second version is given by setting $s = 13$ and $s' = 7$. The latter model is thought to be a fairly accurate representation for nonpolar monatomic molecules, and is used quite extensively in equation of state calculations. The Sutherland model has three adjustable parameters and the general Lennard–Jones model has four; by appropriately choosing these quantities the theoretical values of μ and κ can be made to fit the experimental values over a wide range of temperatures. Sample comparisons are shown in Figures 12–1 and 12–2; extensive results of this type are presented in the book by Hirschfelder, Curtiss, and Bird.

12–2 SPECIAL FLOWS

A somewhat deeper probe into the validity of Boltzmann's equation is provided, in principle, by comparing theory with experiment for particular flow problems. These results often include the determination of transport coefficients (as in the Couette flow and heat flow studied in Chapter 9), but in addition the theory is also required to predict more subtle phenomena such as shock structure and density and temperature

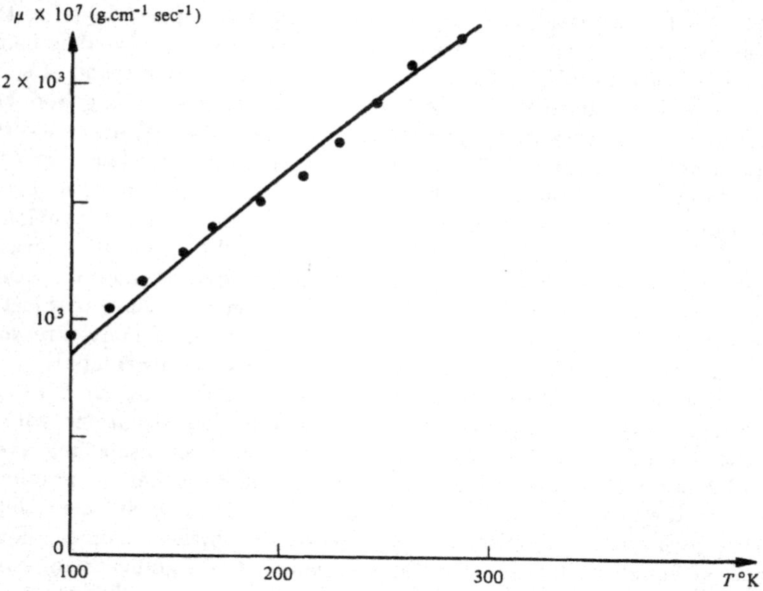

Figure 12–1 Comparison of theory with experimental values for Argon Theoretical curve calculated using a Lennard-Jones 6-12 potential. Data points are from the book by Hirschfelder, Curtiss and Bird.

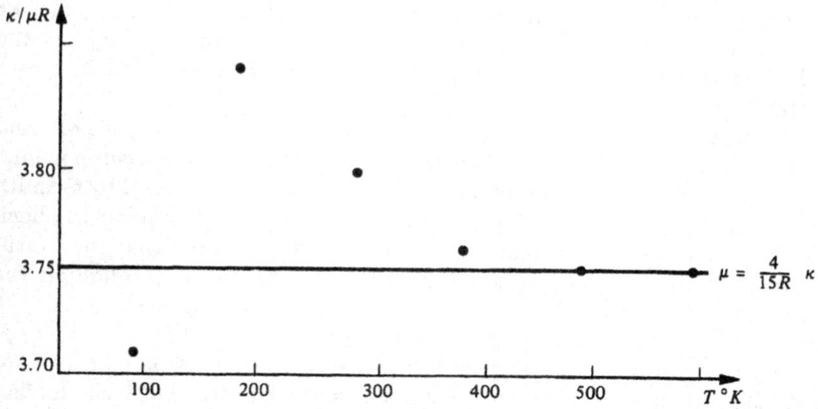

Figure 12–2 Experimental values of $K/\mu R$ for Argon. Data taken from the book by Hirschfelder, Curtiss, and Bird.

profiles. Unfortunately, the state of the art has not yet developed to the point at which we can state with absolute certainty that what is being tested is Boltzmann's equation rather than a particular method of solution. Also, in addition to the possibility of discrepancies arising from the improper choice of a molecular model, we must often also specify the gas-surface interaction in flow problems, which adds another degree of uncertainty to the theoretical results. Nevertheless, certain conclusions can still be drawn by focusing our attention on particular flow problems.

The shock structure problem (Section 9–5) will be accorded a special significance here as it does not involve surface-gas interactions. Also, it is one of the few nonlinear problems which can be treated within the context of the Boltzmann equation rather than a model. Finally, a reasonable amount of reliable experimental data is available for comparison.

Weak and moderate shock waves (Mach number less than 2) can be successfully treated by using the Navier–Stokes equations with a temperature dependent viscosity coefficient. This is because at relatively small Mach numbers the shock wave is fairly thick, so that the gradients through the shock are not large. (Attempts to describe stronger shock waves with either the Burnett equations or the thirteen moment equations, although well motivated, do not lead to any significant improvement of the Navier–Stokes theory. Further, there are no solutions of the thirteen moment equations for $M > 1.65$ or of the Burnett equations for $M > 2.1$.) The description of strong, shock waves, where the Navier–Stokes equations no longer give an accurate description, is thus an example of a problem which is best treated using the methods of kinetic theory. It is important also to point out that only when the Navier–Stokes equations are considered as a consequence of Boltzmann's equation is the necessary temperature dependence of the viscosity coefficient explicit. Neglecting the temperature dependence of viscosity in the Navier–Stokes description of weak shock waves leads to an incorrect (too small) value of the shock thickness.

In the case of strong shock waves, the shock thickness falls off, and the flow is essentially divided into upstream and downstream components. This is just the condition under which we would expect the Mott–Smith bimodal approximation to be particularly valid. Measurements of shock thickness have been made up to $M = 10$, and it is found that the Mott–Smith value compares reasonably well with experiment throughout the strong shock range. This is shown in Figure 12–3. The data for $M < 5$ have been taken from optical reflectivity measurements, while the data for $M > 5$ are from electron beam measurements. A Lennard–Jones 6–12 potential ($s = 13$, $s' = 7$) has been assumed for the molecular model.

A somewhat more demanding test is the comparison of the density and temperature profiles across the shock wave. The available experi-

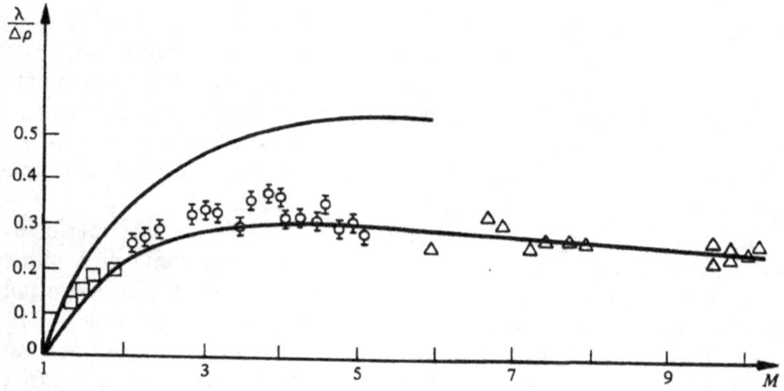

Figure 12–3 Reciprocal shock density thickness for Argon. Lower curve is Mott-Smith calculated value for Lennard-Jones 6-12 potential. Upper curve is Navier-Stokes calculated value. Data points are from Ref. 5 (□), Ref. 6 (Φ), and Ref. 7 (Δ).

mental data here are still limited; however, preliminary studies are in reasonable agreement with the Mott–Smith theory. In this respect we mention that it is possible to simulate the Boltzmann equation description of a shock wave on a computer using Monte Carlo techniques. A particular advantage of such an "experiment" is that the molecular model can be specified, allowing a direct test of various analytical methods of solution. Preliminary results for hard spheres agree quite well with the Mott–Smith solution in the strong shock regime.

Thus, we see that the Mott–Smith solution of Boltzmann's equation predicts results in close agreement with experimental findings in the strong shock regime, $M > 2$. A modification of the Mott–Smith method (Reference 23, Chapter 9) extends this agreement into the moderate shock regime.

Much less can be said about the flow problems considered in Chapter 9 which involve gas-surface interactions. Experimental results for Poiseuille and Couette flows are only available for cylindrical geometry, whereas theoretical results are only available for plane geometry. Certain qualitative comparisons can be made however. Thus, in the case of Poiseuille flow, the experimentally observed minimum in flow rate (as a function of pressure) is reproduced by the theory. In the case of Couette flow, even the crude model considered in Section 9–7 predicted a velocity slip. The velocity profile found in Section 9–7 was linear, a result which is also obtained by treating the problem with the thirteen moment equations. The Couette flow problem has also been solved using the half-range polynomial methods, and here it is found that there is a layer, of thickness less than a mean free path, in which the velocity profile is not linear. Although the presence of this Knudsen layer has not been directly verified by experiment, related experimental results indicate its existence. It would appear, in the light of the scarcity of experimental results, that both the Poiseuille and Couette flows would be of particular importance from the point of view of future computer studies.

In the case of the heat flow studied in Section 9–6, some experimental results are available. Measurements of the density profile in a planear geometry are in good agreement with the four moment half-range polynomial approximation [Equation (9–51)] discussed in Section 9–6. These results were obtained for Argon gas over a wide range of Knudsen numbers. Surprisingly enough, the eight moment approximation compares markedly poorer with experiment. On the other hand, Monte Carlo results compare favorably with the eight moment approximation, and so no definite assessment should probably be made at present.

Finally, a comparatively subtle probe of the Boltzmann equation, for which reliable experimental techniques are becoming established, is provided by light scattering experiments. The recent development of lasers as a diagnostic tool which serves as an intense monochromatic light source is responsible for the progress being made in this area. The connection between experiment and theory is as follows. The light scattered by the density fluctuations in a gas has a measurable frequency distribution which is proportional to $S(\mathbf{K},\omega)$. When the fluctuations are on the order of a mean free path, $S(\mathbf{K},\omega)$ can also be computed by solving the Boltzmann equation as an initial value problem (as described in Section 9–4). A comparison of the measured and calculated values of $S(\mathbf{K},\omega)$ appears to offer a far superior test of the applicability of Boltzmann's equation to rapidly varying phenomena than does a comparison of high-frequency sound propagation results. In the latter case the experiments are very difficult to carry out, and the theory remains subject to a continuing controversy.

Calculations of $S(\mathbf{K},\omega)$ using the generalized BGK equation for both hard spheres and Maxwell molecules have been compared with experimentally obtained results for the scattering of single-mode laser of 6328 Å by xenon at 24.8°C and 780 mm Hg. Both calculations compare very well with experiment, but the hard sphere results are a little superior in describing the finer structure. This is certainly an area which should see a great deal of activity in the coming years.

12–3 THE DISTRIBUTION FUNCTION

The most significant test of the validity of the theory we have developed would be to compare actual solutions of Boltzmann's equation with experimentally determined values. One case where this is actually feasible, and where the necessary experiments have been carried out, is that of thermal equilibrium. As we saw in Chapter 5, one of the consequences of Boltzmann's equation is that it predicts that an isolated system will approach a state of equilibrium in which the distribution function is uniquely determined. The equilibrium distribution function is the same function found by the independent theoretical methods of equilibrium statistical mechanics, and thus certain of its moments, such as the pressure and the heat capacity, are known to compare well with experiment. A comparison of the distribution function itself, however, provides a much more demanding test.

An experiment in which the equilibrium distribution function itself is determined can be carried out using molecular beam techniques. A metal (cesium and bismuth are commonly used) is vaporized in a small oven which is kept at uniform temperature, so that a gas in thermal equilibrium is thus produced. This gas is allowed to effuse through a very narrow slit into an evacuated chamber provided with a second, collimating slit, in which the effusing molecules are acted upon by a deflecting external field. The molecules are then collected on a detector at the end of the chamber which measures the intensity and position of the deflected effusing beam. When gravity serves as the external deflecting force, the vertical intensity distribution in the beam represents the velocity distribution of the vapor molecules. The distribution function thus determined is in very good agreement with the calculated Maxwellian function, with a slight discrepancy being found for very slow molecules.

A more sophisticated technique for measuring distribution functions, involving the measurement of Doppler profiles, has recently been developed. This technique offers the possibility of making measurements on actual flow systems as well as systems in equilibrium. Preliminary results for equilibrium conditions and for a free jet expansion are in good agreement with theory.

12–4 COMPUTER EXPERIMENTS

The past decade has seen the development of several innovative applications of computer techniques in the area of kinetic theory. While the computer is still a long way from taking on the role of a Lagrangian demon,[1] progress in this direction is being made, and in the present section we will briefly consider some of the techniques currently in use.

The general idea of a computer experiment is to simulate the motion

[1] A Lagrangian demon is a member of that branch of the demon family which also includes the more famous Maxwell demon. A Lagrangian demon's powers enable him to solve the equations of motion for all the constituent particles in the universe, and thereby construct all past and future history.

of a system of molecules on a computer by solving the equations of motion for what is hopefully a representative number of particles. At present, computer capabilities limit the size of the systems that can be treated to $O(10^3)$ particles. Thus, in order to minimize the effects of fluctuations (which are inversely proportional to the square root of the sample size, and hence greatly enhanced in a computer simulation), only flows with large pertubations are generally considered. The particular relevance of computer experiments to the theory which we have developed is not in allowing us to test Boltzmann's equation for a given flow, but rather to test the competing approximate solutions of that equation for the flow. Since the interparticle potential (and, if required, the gas-surface interaction) is specified in the computer simulation, direct comparison between theory and experiment is possible.

The most direct method of computer simulation is the molecular dynamics technique introduced by Alder and Wainwright. In this approach the evolution of a system of molecules interacting through some prescribed interparticle potential is followed in a deterministic fashion by explicitly solving the equations of motion on the computer. This method has not yet been applied to the study of flow problems since the computing requirements for a meaningful sized sample are not practically available. This method has been used to demonstrate the equilibration of a randomly prepared system of 100 molecules in a box, each molecule initially having the same speed and randomly oriented velocity vector. At present, however, the use of this technique is limited to the study of dense systems, where it is used to compute the equation of state and correlation functions of dynamical variables for systems in equilibrium.

A considerable reduction of the computing requirements can be effected by computing the collisions in a probabilistic rather than a deterministic manner, and this is the basis for the so-called Monte Carlo method of computer simulation which has been utilized to some extent in treating flow problems in kinetic theory. The method which we will consider here, due to Bird, is actually quite similar to the molecular dynamics technique in that the motion of all of the molecules is considered rather than just that of selected test particles. The first step in the Monte Carlo procedure is to divide the spatial region in which the flow is contained into a large number of cells whose dimensions are small compared with distances over which significant changes in the thermo-fluid variables occur. The sample molecules are then uniformly distributed throughout the physical space and assigned velocities corresponding to a Maxwellian distribution in accordance with some random number selection process. Boundary effects characteristic of the flow being studied are then introduced, producing an unsteady flow which tends, after a long time, to a steady flow.

The evolution of the system is computed using the same assumptions that are implicit in Boltzmann's equation; binary collisions, short-range potentials, and a *stosszahlansatz* are each assumed, and it is in this sense that the method simulates the solution of Boltzmann's equation. In computing the evolution of the system, the free motion of the molecules is decoupled from their motion while interacting by using the following scheme. Collisions taking place in some small interval Δt are computed in each cell, and then each molecule is moved through the distance Δt times its instantaneous velocity.

The collisions themselves are calculated in each cell by determining the collision cross section for randomly selected collision partners taken from the cell using the particles relative velocity and a randomly chosen impact parameter. These two quantities plus the assumed molecular model then serve to determine the post-collisional velocities. Implicit in the random selection of collision partners and impact parameters in each cell is the assumption that the resulting random values generated offer a representative sample of the gas at that particular spatial location.

After each collision in a cell, the time at that cell is advanced $\Delta t_i = (2/N_c)(SnV)^{-1}$, where N_c is the number of molecules occupying the cell, n the number density, and S a collision cross section which depends on the molecular model. Collisions are computed in each cell until

$$\sum_i \Delta t_i = \Delta t$$

at which time each molecule is moved the distance Δt times its instantaneous velocity. A description of the flow is then obtained by sampling the flow properties at selected intervals. In steady flows it is customary to average over several intervals to increase the sample size and reduce the effect of fluctuations.

As has been mentioned in the preceding sections, the Monte Carlo method has been successfully used to describe several flow problems of interest, and this will no doubt be an area of considerable activity for future researchers.

References

A large number of experimental results for transport coefficients for monatomic pure gases, monatomic binary mixtures, polyatomic gases, and even dense gases are given in

1. J. O. Hirschfelder, C. F. Curtiss, and R. B. Bird, *Molecular Theory of Gases and Liquids.* New York: Wiley, 1954.

See also

2. S. Chapman and T. G. Cowling (1952). Note General References.

3. L. Waldmann (1958). Note General References.

4. *Thermal Conductivity*, vol. 2, R. Tye, Ed. New York: Academic Press, 1969.

Experimental results for shock wave structure have been obtained by

5. F. S. Sherman and L. Talbot, in *Rarefied Gas Dynamics*, F. M. Deveine, Ed., New York: Pergamon Press, 1960.

6. M. Linzer and D. Honig, *Phys. Fluids*, vol. 6, p. 1661, 1963.

7. M. Camac, in *Rarefied Gas Dynamics*, vol. 1, J. H. de Leeuw, Ed. New York: Academic Press, 1965.

8. F. Schultz–Grunow and A. Frohn, *ibid.*

Monte Carlo solutions for a shock wave have been given by

9. G. A. Bird, *J. Fluid Mech.*, vol. 30, p. 479, 1967.

10. B. L. Hicks and S. M. Yen, in *Rarefied Gas Dynamics*, vol. 1, L. Trilling and H. Wachman, Eds. New York: Academic Press, 1969.

Experimental results for the heat flow problem have been obtained by

11. W. Teagan and G. Springer, *Phys. Fluids*, vol. 11, p. 497, 1968.

Monte Carlo results for the heat flow problem have been given by

12. J. Haviland and M. Lavin, *Phys. Fluids*, vol. 5, p. 1399, 1962.

13. S. M. Yen and H. J. Schmidt, in *Rarefied Gas Dynamics*, vol. 1, L. Trilling and H. Wachman, Eds. New York: Academic Press, 1969.

Comparison of theory with experiment for laser scattering has been done by

14. M. Nelkin and S. Yip, *Phys. Fluids*, vol. 9, p. 380, 1966.

15. A. Sugawara, S. Yip, and L. Sirovich, *Phys. Rev.*, vol. 168, p. 121, 1968.

Molecular beam techniques for measuring the Maxwellian distribution function were originated by

16. I. Estermann, O. C. Simpson, and O. Stern, *Phys. Rev.*, vol. 71, p. 238, 1947.

Doppler profile measurement techniques are due to

17. E. Muntz, in *Rarefied Gas Dynamics*, vol. 2, J. H. de Leeuw, Ed. New York: Academic Press, 1966.

The first reported computer studies of molecular systems were the molecular dynamics simulations done by

18. B. J. Alder and T. E. Wainwright, in *Proc. Internat'l Symp. Transport Processes in Statistical Mechanics*, I. Prigogine, Ed. New York: Interscience, 1958.

Monte Carlo results for the equilibration problem considered in Reference 18 were obtained by

19. G. Bird, *Phys. Fluids*, vol. 6, p. 1518, 1963.

The Monte Carlo method considered here was proposed by

20. G. Bird, in *Rarefied Gas Dynamics*, vol. 1, L. Trilling and H. Wachman, Eds. New York: Academic Press, 1969.

The Monte Carlo test particle method has been discussed by

21. J. K. Haviland, in *Methods in Computational Physics*, vol. 4, B. J. Alder, Ed. New York: Academic Press, 1965.

22. A. Nordsieck and B. L. Hicks, *Rarefied Gas Dynamics*, vol. 1, C. L. Brindin, Ed. New York: Academic Press, 1965.

general references

The three advanced general references mentioned in the preface are:

1. T. Carleman, *Problems Mathematiques dan la Theorie Cinetique des Gaz.* Uppsala, Sweden: Almquist and Wiksells, 1957.

2. H. Grad, in *Handbuch der Physik,* vol. 12. Berlin: Springer-Verlag, 1958.

3. L. Waldmann, in *Handbuch der Physik,* vol. 12. Berlin: Springer-Verlag, 1958.

Another advanced general reference, which has just been published, is:

4. C. Cercignani, *Mathematical Methods in Kinetic Theory.* New York: Plenum Press, 1969.

In addition to the above, the following is a standard reference for much of the material contained in the first six chapters:

5. S. Chapman and T. G. Cowling, *The Mathematical Theory of Non-Uniform Gases.* Cambridge, England: Cambridge University Press, 1952.

Finally, a short but very lucid account of the theory of Boltzmann's equation is contained in chapters IV, V, and VI of

6. G. E. Uhlenbeck and G. W. Ford, *Lectures in Statistical Mechanics.* Providence, Rhode Island, American Mathematical Society, 1963.

The original expositions of both Maxwell and Boltzman are exceptionally readable, and both are highly recommended:

7. L. Boltzmann, *Lectures on Gas Theory* (English translation by S. Brush). Berkeley, Calif.: University of California Press, 1964.

8. J. C. Maxwell, *The Scientific Papers of J. C. Maxwell.* New York: Dover, 1965.

A complete list of references to the early papers in kinetic theory, and related material, is given in Reference 7. Of particular interest is the memoir on Boltzmann:

9. G. Jaffe, *J. Chem. Educ.*, vol. 29, p. 230, 1952.

subject index

subject index

author index

author index